Weather
and the
Animal World

Also by J. H. Prince

Animals in the Night: Senses in Action After Dark

The Universal Urge: Courtship and Mating Among Animals

Comparative Anatomy of the Eye

The Rabbit in Eye Research

Weather and the Animal World

by J. H. Prince

THOMAS NELSON INC.

NASHVILLE · CAMDEN · NEW YORK

Copyright © 1974 by J. H. Prince

All rights reserved under International and Pan-American Conventions. Published in Nashville, Tennessee, by Thomas Nelson Inc., and simultaneously in Don Mills, Ontario, by Thomas Nelson & Sons (Canada) Limited. Manufactured in the United States of America.

First edition

Library of Congress Cataloging in Publication Data

Prince, J. H.
    Weather and the animal world.

    SUMMARY: Discusses weather and its effects on land and water animals.
    1. Bioclimatology—Juvenile literature.
2. Zoology—Ecology—Juvenile literature.
[1. Weather. 2. Bioclimatology] I. Title.
QH543.P74   591.5′42   74–10270
ISBN 0–8407–6416–2

CONTENTS

Introduction   11

PART ONE THE CREATION OF WEATHER   15

   1/Weather in the Making   17
      ATMOSPHERIC ZONES   17
      THE PARTS PLAYED BY WATER AND LAND   18

   2/Winds   23
      TRADE WINDS   23
      HURRICANES   27

   3/Wind and Waves   31

   4/Pressure Systems   35
      THE REVOLVING WINDS   35
      READING A WEATHER MAP   38

   5/Fronts   43
      COLD FRONTS   43
      WARM FRONTS   43
      OCCLUDED FRONTS   44

PART TWO WEATHER AND ITS EFFECTS ON
        FISH AND WATER BIRDS   49

   6/How Fish Are Affected by
     Temporary Climate Change   51

7/The Effects of Light on Fish	53
PENETRATION OF LIGHT INTO WATER	54
THE EFFECTS OF EXCESSIVE LIGHT	55
8/The Influence of Temperature on Fish	59
TEMPERATURE CONTROL OF FISH GROWTH	59
TEMPERATURE AND FISH HABITS	60
THE RESPONSE TO TEMPERATURE CHANGES	61
9/Understanding Ocean Tides	63
THE CREATION OF TIDES	63
UNUSUAL TIDES	68
RIVER TIDES AND SALINITY	70
10/The Effects of Tides and Salinity on Fish Behavior	73
FISH FEEDING HABITS AND TIDAL CYCLES	73
TIDES IN ESTUARIES	74
RESPONSES TO CHANGES IN SALINITY	74
11/The Effects of Storms on Shore Areas and Estuaries	77
ESTUARY MORTALITY	78
MORTALITY IN FISHING BIRDS	80
THE EFFECT OF STORMS ON MIGRATION THROUGH ESTUARIES	82
RESPONSE TO CHANGES IN ATMOSPHERIC PRESSURE	82
PART THREE HOW LAND ANIMALS ARE AFFECTED	**85**
12/The Effects of Climate Variation on Breeding	87
SEVERE AND PROLONGED WINTERS	87
RESPONSES TO RAIN	90
13/The Effects of Temperature on Breeding	93
ANATOMICAL AND PHYSIOLOGICAL CHANGES	93
EFFECTS ON MATING HABITS AND THE YOUNG	95

THE IMPORTANCE OF INSECTS	95
THE EFFECT OF TEMPERATURE ON THE BEHAVIOR OF BEES	97
14/Special Adaptations	99
HIBERNATION	99
REACTIONS TO HEAT AND ARIDITY	102
THE GARUMA	104
AESTIVATION	105
15/Extreme Weather Conditions and the Reduction of Animal Populations	109
DROUGHT	109
RABBITS ARE NOT INDESTRUCTIBLE	111
DESTRUCTION BY HURRICANES AND TORNADOES	112
FLOODS	113
Index	117

Weather
and the
Animal World

INTRODUCTION

When we wake in the morning and see rain teeming down, or there is a gale blowing in all directions, our spirits usually sag. If such weather persists for days, we may even become depressed. During a long period without rain, a veritable drought, we are less likely to be depressed; yet drought can affect our lives just as much as excessive rain because of its effects on the plants and animals in our environment or on our farms.

We may not feel the effects at once. Nevertheless, when animal breeding sites are destroyed by floods, or drought reduces the insect population that serves as food for small animals, these animals are unable to reproduce normally, and since they, in turn, represent food for larger animals, quite a chain of disaster can follow. In the cities we may not be aware of it, but eventually every adverse effect of this kind has *some* influence on our lives.

One of the reasons we study weather is to reach a better understanding of its effects on the environment, on animals, and, through them, on ourselves.

The farmer never has to be reminded that if seasonal rain is late in arriving or fails to arrive at all, his feed grain will be stunted or will mature late, and as a result of this his cattle will lose weight, his cows will give less milk, his sheep will provide poor-quality wool, pests and diseases will increase on his property, and vermin will move into his living area. His whole life cycle will be disturbed and his income reduced.

Such problems are not confined to farms. They reach out to the entire animal kingdom. Fewer field mice, for instance, mean less food for owls or foxes. The foxes then find the farmer's chickens more attractive; the terrified hens produce fewer eggs, and so it goes.

Before we can study the effects of weather and climate on animals, we must understand how weather is created, and why it sometimes causes catastrophic changes in our lives through changes in the animal world. The two environmental areas of animal life, land and water, often respond differently to unusual weather. While marine breeding and living areas can be affected or even destroyed just like land areas, the effects are not always so long lasting, unless they are compounded by man-made pollution.

It is possible that we know more about how weather affects marine life and its environment than we know about how it affects the land, because of systematic observations made by generations of professional fishermen and, in more recent times, by increasing numbers of skindiving scientists, who can evaluate the otherwise unseen effects of storms, seaquakes, river runoff from great land storms, hurricanes, and unusual temperature changes.

Man has always been concerned about domestic animals because, like fish, they have always been important to his diet. However, our knowledge of the effects of weather on other land animals is probably more recent. Until the last thirty or so years, many species of wild animals were hunted, either for sport or as pests, without a thought for their ecological significance, so the effect of weather and climate on their population was seldom considered.

With the knowledge we are now gaining, we have discovered some very interesting and unexpected factors. For instance, it has been found that many species of fish *must*

INTRODUCTION

stay within a critical temperature range, or they will die. For most species this range is only six or seven degrees, and fish avoid or leave water that rises above or falls below their range. Next to pollution, temperature change is one of the most important environmental factors for fish, and such change is related to the surface weather as well as to currents.

A knowledge of the effects of weather on water temperature is therefore very important to those who farm the sea, lakes, and rivers. To give a simple example, trout need a relatively low temperature and high oxygen pressure, so when the temperature becomes unusually high, they stay deep in pools or running gullies. After it has rained and when the air is cool, they will rise to feed near the surface, because the rain has cooled the surface water to a level of their tolerance. If they cannot get to cool water when the weather is unseasonably hot, they may die.

Quite a number of other fish react in the same way. In fact, so critical is this temperature factor that a study of the temperature patterns of the present season will make it possible to forecast the average size and weight of fish as well as the population number of a species in the following season.

While the atmospheric pressure would seem unlikely to affect fish, because of their natural ability to adapt to changing water pressure as they move up or down to deeper water, many people feel sure that it does. Wind and rain certainly affect fish—wind because it increases the oxygen in the surface layers of the water, rain because it cools these layers. Rain also dilutes the saline in estuaries and brings destructive debris and sediment down to the ocean, making it murky and unattractive to many species of fish, sometimes even destroying fish in large numbers.

Because we must necessarily separate the description of how weather is created from its effects on particular animals, some of the chapters in this book are devoted entirely to weather, others to the effects of weather on the animal world. It should be easy to integrate them as they are read.

The two great environmental areas, water and land, are kept separate, because although what happens on land often affects lakes, rivers, and oceans, and vice versa, it will be more comprehensible to the reader if we keep them separate.

PART ONE
THE CREATION OF WEATHER

1 / Weather in the Making

EVERYTHING HAS A SMALL BEGINNING, and one can say that even the most destructive storm must have developed from what was originally a light breeze. In order to understand weather changes and their destructive potential, we must first understand how particular conditions are created, the influence of the sun and the moon, and where the elements of the changes originate. We can deal with only some of the theory of weather here, and that may seem quite simple, considering all our present knowledge, but except for the tropics, even the experts can be proved wrong in their forecasts by sudden and unexpected changes resulting from the most trivial unknown factors—a slight change of wind direction, a change of temperature brought from some far-off place, or something happening in the upper atmosphere, especially during the night.

ATMOSPHERIC ZONES

The outer limit of the atmosphere is probably a thousand miles above the earth at the most; we are not quite sure exactly how far it does extend. From an altitude of about fifty miles to about five hundred miles is a region known as the *thermosphere*. Great heat is encountered in the upper part of this zone, with temperatures rising as high as 4,000° F. (2,260° C.). In the lower portion of the thermosphere, at an altitude of about eighty miles, the temperature decreases sharply. The coldest zone of the atmosphere is between fifty and sixty miles high, where the temperature drops to as low as −130° F. (−90° C.). Below the temperature, at a height of about fifty miles, comes the *mesophere*, and here the temperatures rise again. Between twenty and forty miles up they hover around 30° F. (−1.1° C.). Nowhere in these regions is there any weather.

Below the mesosphere is the *stratosphere*, which is weather-free but is not necessarily free of wind. In the higher regions of the stratosphere, the temperature falls sharply again, but in the lower stratosphere it becomes very uniform, and this region of uniform temperature ends at a height varying from ten miles at the equator to five miles at the North and South poles.

Below the stratosphere is the *troposphere*, the layer of the atmosphere in which we live. It contains water vapor in addition to the gases that make up the air we breathe. It is this mixture of water and gases that gives the troposphere its weather, which is caused by the heating of water and earth by the sun, and by the uneven heating and cooling of air with water vapor in it.

THE PARTS PLAYED BY WATER AND LAND

Water and land absorb heat from the sun at different rates. Water absorbs it more slowly but also releases it more slowly, so the land heats up rapidly during the day and cools off rapidly during the night. This means that temperature variations are greater over land than they are over water.

Heated land warms the air above it. This air rises high into the atmosphere, pushed up by the cooler air that flows in from over the oceans to take its place. The result is a sea breeze, or onshore wind. As the warm air rises, it takes moisture with it.

The ocean retains much of the heat it picks up during the day, and at night this makes the air above it warmer than the

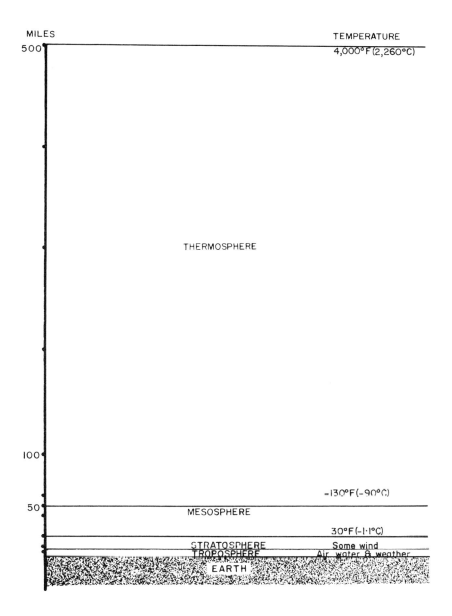

Figure 1 The earth's air-filled envelope, the troposphere, is a small part of the entire atmosphere, which has temperatures ranging from −130° F. to 4,000° F. (−90° C. to 2,260° C.).

air over the land. This warmer air over the sea is then pushed upward by cooler air from the land, and it takes still more moisture with it. The cooler air that flows from over the land is the familiar land breeze.

These light breezes can hardly be classed as weather, but they are in fact miniature weather-forming patterns, and

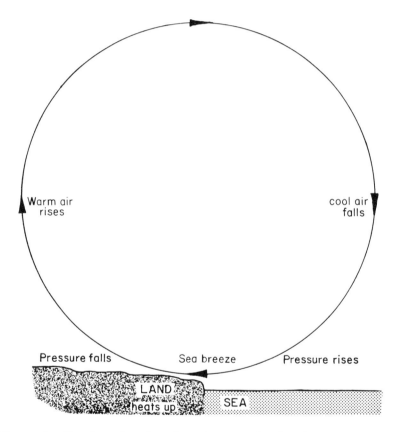

Figure 2 During the day the air over the land warms up and rises, and the air pressure goes down. Cooler air moves in from the sea and takes the place of the warm air. Then warmer air from the higher regions moves toward the sea, where the pressure is greater. The air cools as it descends once more to sea level.

this is demonstrated by the onshore breezes we have just mentioned, which are sometimes so intense that they build up into squalls. A breeze, a squall, a storm—it is only a matter of degree or intensity.

Ordinarily, whirling columns of air reach heights of only a few hundred feet, but over exceedingly hot land they may sometimes extend half a mile upward. These convection currents are not always visible and are usually localized anyway, but they are numerous enough to play a part in the overall circulation of the atmosphere. As well as a rotary motion, they have strong up-currents, which carry heat from the ground to much higher levels.

Clouds are formed above such currents, and compensating downward currents occur in and below the drier spaces between these clouds, especially in tropical areas. This upward circulation is one of the principal ways in which water is carried to the upper atmosphere.

We see from all this that the creation of breezes is dependent on temperature changes, and that the formation of clouds depends on rising temperature and the presence of water. But breezes vary in intensity, and there are many kinds of clouds. When the intensity of a breeze passes a certain velocity, the breeze becomes what most of us call wind, and winds not only create weather but also carry it from place to place, thus creating changes in weather as well. For example, because of favorable winds, Machupicchu, a high valley in the Peruvian Andes, is warmer in winter than in summer.

2 / Winds

THE MOST CONSTANT WINDS are found over the tropical oceans, where there is very little temperature change either during the day or during the year. There is just a steady and constant displacement of warming air by cooler air from other latitudes.

TRADE WINDS

Around the equator a belt of easterly winds nearly encircles the earth. These are the trade winds, the steadiest of all earthly air currents. They keep the climate much more stable than do the winds in temperate zones, and apart from the storms that sometimes build up, they do more to make the tropics habitable than anything else.

Air currents from the northern and southern hemispheres converge and rise in the tropical zone between the tropics of Cancer and Capricorn (see Figure 4), and in spite of the general stability of the area, these rising currents often produce rather heavy cumulus clouds which in turn give birth to thunderstorms.

Small changes in air circulation here can have a considerable effect on the weather. Great cloud towers may develop and persist for many hours; hurricanes and cyclones with cumulonimbus clouds (see Figure 5) and violent convections may develop enough energy to take them right out of the tropical zone.

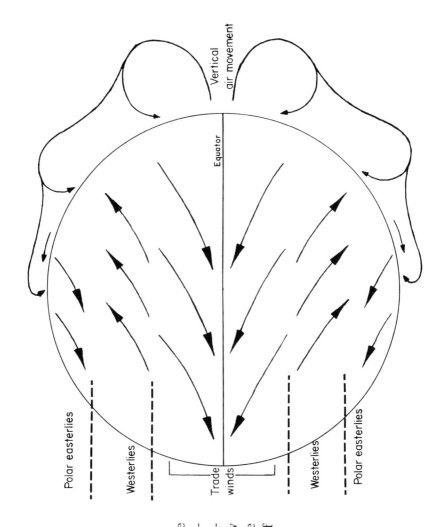

Figure 3 The tracks of the prevailing winds on the surface of the earth. The diagram attempts to show how these winds are related to the great upward movement of air in tropical regions.

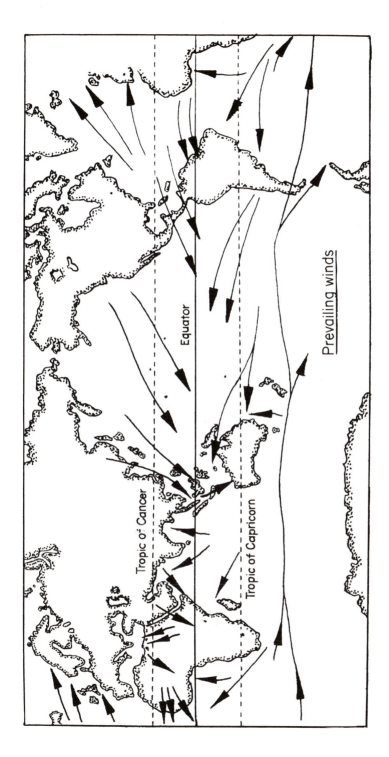

Figure 4 The general direction of the earth's air-current movements in relation to the continents.

Figure 5 When cloud masses rise high, they are carried up by powerful currents of warm air, and these are frequently followed by storms. The bottom picture shows a typical cumulonimbus cloud mass. In the top two pictures, cumulonimbus clouds are in the process of formation.

Although heat and moisture usually move vertically over the tropical oceans, they travel more horizontally farther north and south, with no clearly defined line between the two areas. In fact, these currents are part of a continuous system that maintains the heat balance of the atmosphere, with the warm air rising in the tropical regions as cooler air in the form of wind from the other areas displaces it.

At a height of about thirty thousand feet, well above the other systems already described, are the westerlies in the middle latitudes. These winds form a belt that encircles the earth. They affect the weather patterns in the lower zones of our atmosphere and help to make weather prediction an uncertain science.

In normal wind patterns, air coming off the mainland is likely to be dry; air coming from polar regions naturally is cold; and air from equatorial regions is warm and moist, so that its humidity is high. Because high-pressure belts tend to move toward the equator in winter and away from it in summer, scientists can predict to some extent the nature of the winds in those regions.

HURRICANES

In coastal areas winds from the sea can produce rain at almost any time, but such rain is usually short-lived. While gales can occur in or around depressions, or low-pressure areas, the really violent weather patterns that we call hurricanes, typhoons, and cyclones all develop over water, and warm, moist air is their main ingredient. They draw it in and condense it in huge quantities. Such storms are self-feeding. As water vapor is converted to liquid water in clouds and then descends as rain, it releases vast amounts of heat, which in turn causes an inrush of cooler air from other regions and forces the warm air up again.

All this creates strong winds which whip up spray from the ocean, and so make yet more water vapor available to

27

produce more clouds and more heavy rain. That is why we say these storms are self-feeding. As soon as they pass over land, however, they die gradually because of the sudden lack of moist air.

Hurricanes form over water temperatures of 82° F. (27.8° C.), but not in Atlantic waters below the equator, where the water is exceedingly deep, and thus not so warm.

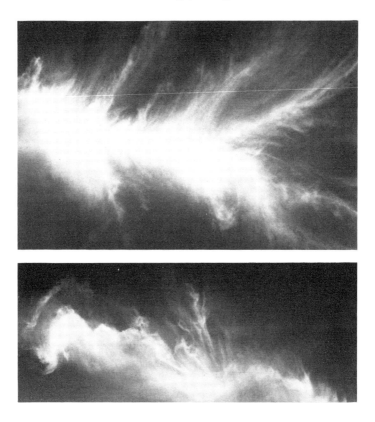

Figure 6 Clouds being torn in all directions by high winds. The cloud at the top was at a considerable altitude and preceded just a mild storm. When a hurricane is arriving, such shredding is seen at the edge of a cloud mass rising from the horizon, and the shreds curl over more, as shown below.

The first warning of a hurricane for those on land is a change in the behavior of shore waves. They may reduce in number per minute by 66 to 80 percent, a day or even several days before the storm arrives. Rooster-tail (veil) clouds, an appearance brought about by wind tearing clouds to shreds, are the first sign of the hurricane's arrival when they are seen rising from the horizon (see Figure 6).

Hurricane wind speeds can vary from seventy-five to two hundred miles per hour. In one day the heat energy released can be equal to 8,000 million tons of explosive, and a hurricane can last from eight to twelve days; some have been known to last a month. The waves these storms create may be as high as thirty-five feet; much higher ones have been reported from time to time.

Into the same category as hurricanes fall the typhoons of the China Sea, which if anything, may be even worse, for their winds have been reported to exceed 250 miles per hour at times.

3 / Wind and Waves

AN ONSHORE WIND, or a wind coming in from the sea, causes the sea to pile up before it. As a result, the sea reacts to rocks, shallows, and coastal irregularities and carries microscopic life, small fish, and other surface-swimming prey in close to shore. An onshore wind will also clear the water sometimes. Offshore winds have the opposite effect. They flatten the sea close to the shore, leaving it relatively smooth, but they raise choppy water farther out.

Local winds have very little to do with swell, or long low waves, but they do control the height of waves. These *cycloidal* waves roll forward before the wind, coil over, and eventually break in surf on the shore. The water itself does not move forward unless it is carried by a current; it is only the wave motion that is moved forward by the wind. If this were not so, it would be impossible to swim against waves.

A light breeze of up to six knots is not likely to do more than ripple the water's surface. A gentle breeze of up to ten knots will create small wavelets. A moderate breeze of up to sixteen knots will begin to raise real waves two to four feet high. Once the wind velocity exceeds twenty knots, the seas rise to eight feet or more, until, at a wind velocity of twenty-seven knots, they are considered to be rough, with waves up to twelve feet high. This is really a gale. When the velocity is at forty-five knots, the waves can be thirty feet high—or even higher, it is said. We can now see how wind,

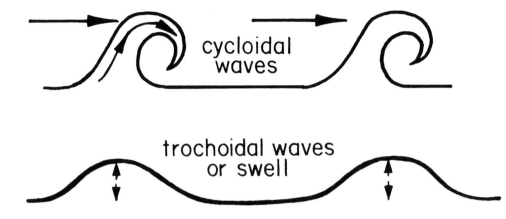

Figure 7 Waves are created by the surface pressure of the wind (large arrows), and although they move forward before the wind, the water surface itself does not move forward. If it did, it would be impossible to swim against waves. Swell is a vertical movement of the water, and it also moves forward away from a distant storm center.

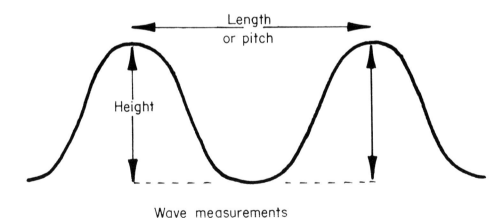

Wave measurements

Figure 8 The height of a wave is measured from the crest, or peak, to the trough. The length of a wave is the same whether it is measured from crest to crest or from trough to trough.

which is created by changes in temperature, can build up into the destructive storms that destroy so much of our environment and animal life when they occur.

Close to shore the effects of high winds are worse than they are at sea, because waves react to both the sea bottom and the shoreline. They begin to do so when the depth is less than 250 feet, and they certainly do so when the depth is less than 50 feet.

Swell, or *trochoidal* waves, is really the motion surviving from cycloidal waves made by a storm some distance away. There are considerable gaps between the crests of swell, but these crests get closer together as they approach land because of the resistance the land offers to their forward movement. Eventually, they break over into surf, adding to the tumultuousness of the coastal water.

When wind injects momentum and energy into the sea's surface, creating waves and currents, much of this energy is transferred from the surface to the deeper layers solely by the motions of the water particles themselves. Certainly, the deeper the water, the longer the waves can become and the faster they can move.

4 / Pressure Systems

COOL AIR MASSES DO NOT RISE. In fact, they may actually be falling at any particular time. This is because they are heavier than warmer air and therefore exert a downward pressure. This pressure will make the barometer rise. Cold and warm air masses do not mix, but displace each other, and this means that they also form into separate systems that are not only independent of each other, but alternate with each other. This is shown in Figure 9.

THE REVOLVING WINDS

High-pressure systems are known as *anticyclones,* and low-pressure systems as *cyclones,* or depressions. These cyclones should not be confused with the cyclones described elsewhere as violent storms, so we will call them depressions or low-pressure systems. Both high- and low-pressure systems revolve, and so create wind currents of varying velocities.

The winds moving around a high-pressure system go in a clockwise direction in the Northern Hemisphere and in a counterclockwise direction in the Southern Hemisphere. Similarly, those moving around a low-pressure system go counterclockwise in the Northern Hemisphere and clockwise in the Southern Hemisphere. This means that when low- and high-pressure systems lie adjacent to each other, their winds

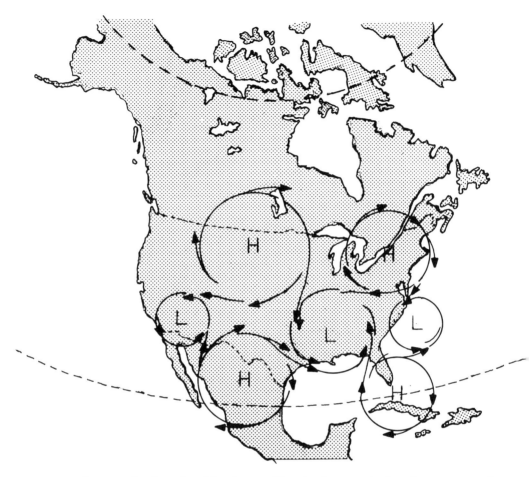

Figure 9 In the Northern Hemisphere winds always travel clockwise around a high-pressure system and counterclockwise around a low-pressure system. They must obviously work together when they are adjacent to each other. This diagram of a theoretical situation shows roughly how this occurs, but it is not necessarily typical of the United States. Winds flow from high pressure to low.

merge, as in Figure 9. When one system is replacing the other, the barometer registers changing pressure accordingly.

It is possible to think of areas of high pressure and low pressure as peaks and troughs, or mountains and valleys of air, that move from place to place. Low-pressure systems

PRESSURE SYSTEMS

Figure 10 In this weather map, the wind revolving around the high-pressure cell on the west coast of Canada would be cool and moist, having passed over the ocean. The wind around the high-pressure cell in the center of the continent should have shed much of its moisture on the land. The wind on the east coast would become moist again as it strikes the shore. Again, this map is theoretical and does not necessarily represent a real weather pattern.

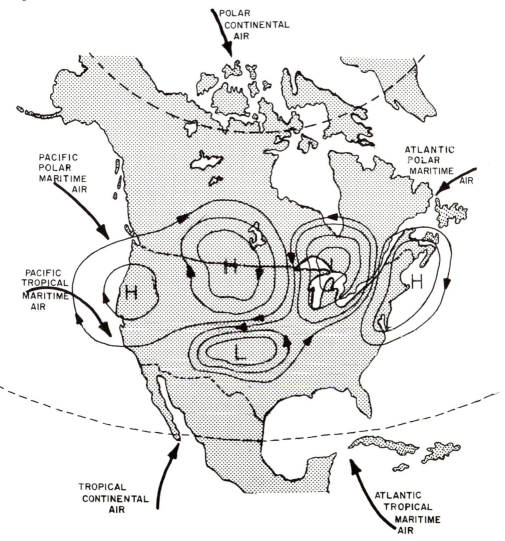

usually move faster than high ones. Sometimes cells of high pressure remain stationary for days, while low systems and fronts pass around them.

Depressions can appear just anywhere overnight, or arrive from other places, and this can turn fine weather into rain quite unexpectedly. The barometer indicates what these high and low systems are doing. If it falls, the assumption is that either a high-pressure system is moving away from the area or a low system is moving into it, and since one system naturally gives way to the other, the decreasing pressure often means both. The reverse is taking place when the barometer rises.

READING A WEATHER MAP

The lines depicting low- and high-pressure systems on a weather map are arrived at when barometer readings are made in many places at the same hour. Those that are identical (after the height above sea level is compensated for) are joined together to form contours on a map, as in Figure 10. These lines are called *isobars,* and they are usually drawn at intervals so that there is four millibars' difference in pressure between them.

When the readings are lowest in the center of a number of encircling isobars, this center is a low-pressure area, or a depression. As the wind flows around this area in a counterclockwise direction in the Northern Hemisphere, it is deflected slightly inward toward the center below heights of two thousand feet.

In high-pressure systems the isobars are highest in the center, and the winds that flow around them have a slight outward deflection from the center. When isobars are close together on a map, it means that the pressure increases or decreases rapidly toward the center, and this indicates high winds; the farther apart these lines are, the lighter are the winds.

PRESSURE SYSTEMS

When a high-pressure area bulges out at a particular spot, this spot is called a *ridge,* and it can be seen clearly in the isobars. When the bulge is from a low-pressure area, it is called a *trough.*

If a ridge of high pressure forces its way between two low-pressure areas, the barometer will rise and then fall again as the systems pass over a given place. At the same time there will be a brief change of weather from warm to cool and back to warm, or from wet to fine and back to wet again.

When a trough passes between two areas of high pressure, it will bring unexpected squalls and high winds; as it passes, the weather will clear again. Troughs almost always bring rain or showers, but they are temporary, lasting only as long as it takes for the trough to pass.

Bad weather tends to be associated with depressions, or low-pressure systems, and fine weather with high-pressure systems, but although this is true much of the time, it is not infallible. Much depends on the direction of the winds brought by a system. If they have passed over the ocean, either system can bring in rain. Conversely, there may be an absence of rain if the winds have not picked up any water on their way.

If the barometer falls rapidly, any storm that accompanies the low-pressure system may be quite short. But if the decrease is slow and prolonged, a long, severe storm could be approaching. However, in that case, there are other indications too. For instance, many seabirds may be seen close to shore or sheltering on the land.

Rising pressure and decreasing temperature will reduce wind, but if the pressure was rather below normal before it started to rise, squalls and rain will arrive first. Any rapid rise in the barometric pressure does not necessarily mean the end of bad weather; it often means more wind and unstable conditions. Wet, unstable conditions will also follow

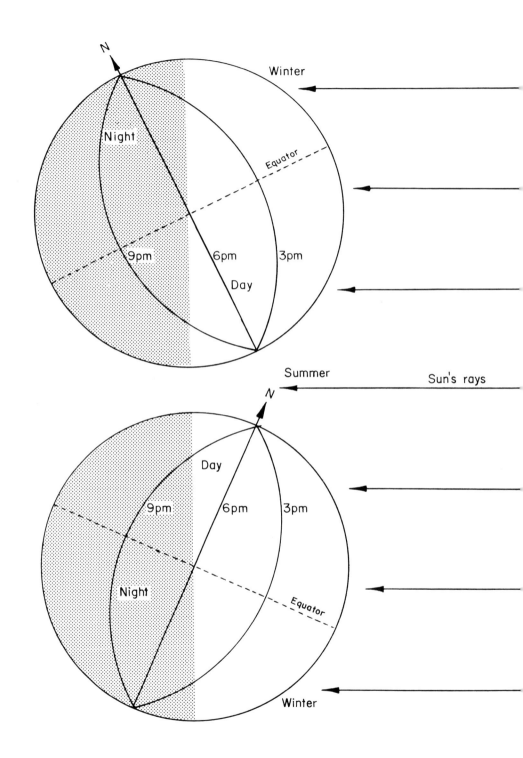

PRESSURE SYSTEMS

fine weather when the barometer falls slowly. Rapid decreases certainly precede rain and storms, a possible change in wind direction, a cold front, or squalls—at any rate, bad weather.

Figure 11 Pressure systems retain their positions more or less relative to the sun, and therefore move north in the summer and south in the winter in the Northern Hemisphere. This is because as the earth tilts its axis relative to the sun, the two hemispheres assume their seasonal day length. In the top picture, the southern hemisphere is having long days and short nights, and pressure systems are farther south, over Australia; in the lower picture, it is summer in the north. In the region of the equator, day length remains constant.

5 / Fronts

AIR MASSES OF DIFFERENT TEMPERATURES and densities do not mix easily; so when they meet each other, they form a *front*. There are three kinds of front: a warm, a cold, and an occluded front. All three kinds produce weather changes and invariably bring some rain.

COLD FRONTS

When a cool or cold air mass from a polar ocean moves toward a warm air mass from the tropics or subtropics and forces its way underneath the warmer air, the latter is pushed up and produces giant cumulonimbus clouds, which then fall as rain. This is the pattern of a cold front. The arrival of this denser, cooler air also brings a rise in pressure, which shows on a barometer.

When the woolly cumulus clouds of a cold front first appear, they are low and dark. Then giant cumulonimbus masses build up into a storm. The cumulus clouds that follow are much higher, and this indicates a fine spell after the storm.

In the Northern Hemisphere, of course, cold fronts come from the north and sometimes the west.

WARM FRONTS

A warm front usually moves more slowly than a cold one, and it can be overtaken by a cold mass. It is due to a body of warm air moving toward, or being overtaken by, a cold air mass and sliding up over the latter as a result of its lightness and easier displacement. The approach of a warm front is seen in high, wispy, filamentlike cirrus clouds, which are followed by a more sheetlike cirrostratus overcast. This overcast later drops to a lower level and becomes caught up with winds that speed it along in the form of what is sometimes called scudders. These winds bring rain in the massive cloud formations that follow.

In many places warm fronts do not always occur at ground level. In the Northern Hemisphere they naturally approach from a southerly direction, because the warm air ordinarily comes from the tropical regions, just as colder air comes from polar regions.

OCCLUDED FRONTS

An occluded front is a very complex pattern in which there are three different air masses, a warm one trapped between two colder ones that force it upward until it forms cloud masses and rain. Whatever the kind of front, therefore, rain is always a possibility.

The kind of front approaching can be recognized in two ways: first by the cloud formation and then by the direction from which it is coming. It is also possible to judge the time that can elapse before its arrival. A cold front comes with

Figure 12 These three diagrams show the cloud patterns associated with the three kinds of fronts. Cumulus clouds always precede cumulonimbus, which in turn are followed by stratocumulus. Similarly, cirrus clouds are followed by cirrostratus, then altostratus, and finally nimbostratus. The top and bottom pictures show cold air pushing its way beneath warm air, but in the center picture of a warm front, the warm air is riding up over the cold.

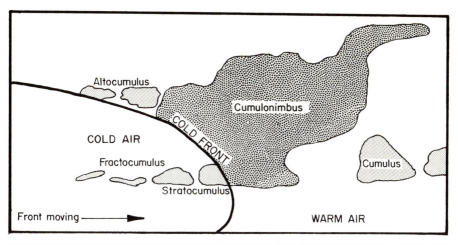
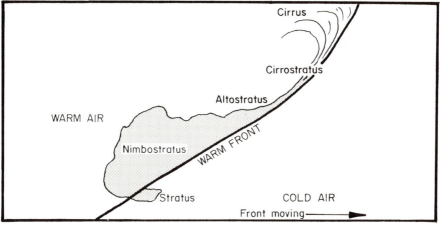
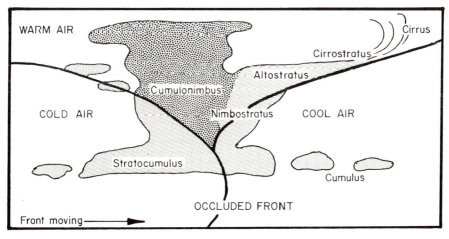

low, dark, woolly cumulus clouds. A warm front comes with high, wispy cirrus clouds, which eventually turn into an overcast sky. The time element can be judged by the speed of the clouds.

We get four basic signs from clouds, and they should be considered when judging the weather. First, when cloud masses are increasing, the weather may be worsening. Second, when clouds begin to move faster, a storm may be approaching. Third, a storm could be very close if two or more cloud formations on different levels are moving in different directions. Fourth, when heavy clouds are approaching from exactly the same direction as the surface wind, it can be assumed that any coming storm will hit from exactly that direction.

Figure 13 Clouds that are moving fast (from right to left in this photograph) and are increasing in volume are often the messengers of an approaching storm, certainly of a worsening weather condition.

FRONTS

It is difficult to think of weather without giving attention to clouds, and clouds would require a large chapter all to themselves; but apart from reducing light and heat, they do not have any direct effect on animals. Their importance for us is in their involvement with storms, which do affect animals considerably.

PART TWO
WEATHER AND ITS EFFECTS
ON FISH AND WATER BIRDS

6 / How Fish are Affected by Temporary Climate Change

CLIMATE AND WEATHER are not exactly the same thing, although they are certainly related to each other. Climate and the seasons are expected to be fairly constant, whereas short-term weather is forever changeable, as everyone knows. Nevertheless, climate does change temporarily now and then, and right now man seems to be doing an excellent job of trying to change it permanently with a blanket of air pollution. Even without ocean pollution, this will in time affect the marine population as well as the animals on land.

Any climate variation from the normal which affects the length of a winter or creates an increase or decrease in the quantity of food available in an area will have a direct effect on the breeding habits of many fish. Among such variations are higher and lower than usual temperatures, higher and lower than average rainfall, and an increase or decrease in the number of cloudy days in a year.

Variables of this kind affect the amount of food available to many species. In addition, interference with day temperature and brightness will upset the balance and production of sex hormones in surface-dwelling fish and invertebrates. Day length and the amount of sun triggers hormones (certain body chemicals manufactured by glands) that stimulate breeding in many creatures both on land and in water, and this can be disrupted by a delayed spring. In fact, varying the day length artificially in experiments with non-tropical species of fish has produced some interesting results in which their whole breeding cycles have been upset.

A long winter may also mean a short summer and so a shorter breeding season for creatures that make up the main diet of fish which, in their turn, breed less because of limited food supplies. When spring rain or the melting of snow on mountains is unduly delayed, breeding ponds and streams used by freshwater fish prove inadequate. Excessively wet winters, on the other hand, may wash away the breeding areas entirely.

Temperature changes in shallow waters at a particular time of year can destroy eggs and kill young fry. In the same way, severe winters that produce lower than usual water temperatures can kill a great many adult fish and crustaceans which cannot move temporarily to warmer water. This not only means fewer fish in following years, but it also means a lower breeding rate the following summer, and so the effects may last for considerably longer than one year.

It is not so with all fish, of course. Some are more adaptable than others. For instance, some species of Pacific shallow-water fish are more tolerant of lower than usual temperatures than closely related Atlantic species. The latter, on the other hand, are more tolerant of higher than usual temperatures.

The effects of climate variations apply less to fish that can choose their location anywhere in the ocean because they can simply follow the temperatures they need in shoals, spawning as they go.

7 / The Effects of Light on Fish

THE EFFECTS OF WEATHER on fish can be indirect through other organisms affected by the weather; organisms on which the fish depend for food. Fish respond more to the movements of their preferred food and to temperature than to anything else. If we have a knowledge of their food and temperature requirements, we can anticipate where fish will be, or where they will go, and when they will be there in any season.

Unless a storm is really violent, it disturbs only the surface layers of the ocean. Below this, calm reigns. The currents are constant in their course, and only two things are likely to change. One is the amount of light, and the other, the surface temperature. The amount of light changes with the state of the sky.

The temperature well below the surface changes only with the variations the currents may bring. Such variations are not very frequent, however, because currents on the whole retain a fairly constant temperature as well as direction when they are not on the surface.

Thus only the surface layers change to any extent, since they are affected by changes in air temperature caused by solar heat, wind, or cooling rain. With increased wind the surface water becomes more generously oxygenated, but that is usually only important in small bodies of water such as lakes and rivers.

Fish find their food through vision, smell, taste, or a combination of a primitive hearing sense and the lateral line, which is sensitive to vibrations. A knowledge of which senses are used by which fish is just as important as a knowledge of how weather influences the fish, because the two often work together.

Among fish living in rivers and lakes that are not deep, vision is usually the predominant sense, except in the bottom grovelers. When the water is clouded with silt after heavy rain or storms, only the fish that use smell and taste to detect their food are likely to be active, especially in the lower reaches of rivers. Those that use vision must either wait for clearer conditions or move to other areas. Rough water may also drive fish to other areas or to deeper water. The deeper waters around a coast usually remain clearer than the shallows, so fish using vision can still feed here, where their prey is silhouetted against the light from the sky. Fish using smell or taste, such as catfish with barbels on their jaw, will usually detect food no matter what the conditions.

PENETRATION OF LIGHT INTO WATER

When the sky is overcast, light penetration into the water is greatly reduced. This may cause certain fish to move nearer to the surface. Some species, however, do exactly the opposite and move to lower regions when the weather is overcast.

As light passes into water, it is rapidly absorbed, and the extent to which this absorption takes place is related to the amount of solid material suspended in the water. Even when water is relatively clear, it has absorbed 50 percent of the light entering it by the time the light has penetrated to a depth of six feet. Almost half of what is left is then absorbed in the next six feet.

THE EFFECTS OF LIGHT ON FISH

A long spell of calm weather and an absence of sedimentation clears water considerably, however—so much so, in fact, that the absorption of light in one place can be up to twenty times as much as in another. The clarity of water can change very dramatically in an area where there is runoff from the land after rain.

THE EFFECTS OF EXCESSIVE LIGHT

It has long been known that prolonged and uninterrupted periods of sunshine reduce the population levels of some fish. Many species must have cloudy weather for spawning, because if there is a sudden excess of sunlight, the young fish may become sick and die. They cannot live in the increased surface temperature. Thus an excess of fine weather can be just as harmful to some species as dull weather can be to others.

In the oceans adult fish can escape persistent sunlight and rising temperatures merely by going a little deeper. But with a few exceptions, fry and young fish must remain near the surface, because that is where their microscopic food is. It is also the place where the dangerous infrared and some of the ultraviolet light rays are absorbed.

Many fish eggs and fry are extremely sensitive to blue and violet wavelengths, and unless they are well shielded by plants, gravel, or rocks, their mortality rate is likely to be almost 100 percent. That is why many river and lake fish spawn in either cloudy or rainy weather or when there is plenty of algae in the water to absorb the harmful light rays. Many of the salmon and trout spawn right into gravel for the same reason. Too much light can be so dangerous for some river and lake fish that their populations will be drastically reduced for long periods after a year or two of continuous sunshine with very little cloudy or rainy weather.

Very few people may realize the connection between a

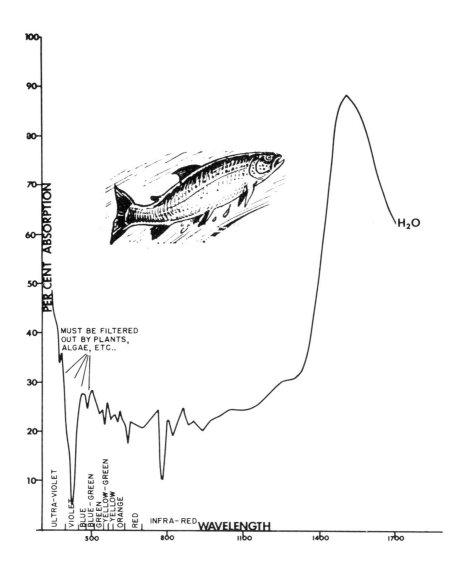

Figure 14 Some wavelengths of light are not absorbed to nearly the same extent as others when they enter water. This graph shows that up to 90 percent of some of the infrared wavelengths are absorbed as they enter the water. The harmful ultraviolet rays, which merge into visible violet wavelengths, are absorbed much less and therefore penetrate farther, with only about 5 percent of the violet being absorbed at the surface.

season of excessive sunlight, with its deadly effects on young fish, and later poor fish harvests. That is because the fish that are born in one year are never harvested until the following or even later years, so the effect of high egg and fry destruction by sunlight and heat is not felt for at least a year, when there may appear to be no reason for it. If a species of fish takes three years to mature, for instance, egg and fry destruction in 1974 will produce a poor harvest of adult fish in 1977.

Experiments with artificial sunlight have now confirmed beyond dispute that weather can be used to forecast this kind of fry mortality and the consequent shortage of adult fish at a later date. Even the extent of the shortage can be estimated.

If, however, an area is well supplied with light-shielding plant life and rocks or even overhanging cliffs or trees, the impact of unseasonal or excessive sunny weather will be reduced.

8 / The Influence of Temperature on Fish

TEMPERATURE CONTROL OF FISH GROWTH

Fish which can migrate well offshore in winter and return close to shore when the temperature around the coast warms up in the spring will delay their return if the winter is long and severe. If their breeding must take place in estuaries, their size, as well as their numbers during following summers, will be affected. This is because the actual growth rate of fish is related to temperature as well as to the length of the summer, and since they are breeding later than usual, their period of growth will be shorter.

Professional fishermen in cold climates can forecast the weight of their likely summer catches by watching the winter temperatures. They know that temperature controls the speed of growth in fish and that lower temperatures will also slow up weight increases. Continuous measurement of winter temperatures even makes it possible to predict the time fish will move into a particular area in certain parts of the world.

In some fish the sex hormones that stimulate breeding can be produced only when lengthening days accompany increased temperatures. This can complicate the cycle, because quite often the days will lengthen while the temperature remains below the required level. Apparently, the effects of abnormal temperature extend even to the embryonic development of fish, for their vertebrae, or spinal bones, and fin rays vary in number with the temperature at the time of incubation. The same thing happens to some amphibians found on land.

TEMPERATURE AND FISH HABITS

Fish always have their chosen feeding times, but temperature plays an important part in those feeding times, because it invariably controls the appearance of the fish's live food. Sometimes fish find the most suitable temperature at the boundary line of two currents, each of which provides a different temperature zone. The fish move back and forth over the boundary, so they can always find the preferred temperature for their immediate needs. Of course, such temperature changes amount to no more than a few degrees, and so the sudden appearance or disappearance of fish in an area may be due to a temperature change that is very slight.

When the temperature increases, the speed of body metabolism, the intake and use of oxygen, and the elimination of the body's waste products increase. As a result, hunger and breeding inclination increase in certain species. A drop in temperature produces less activity. So if for any reason, such as a change in weather, certain fish are unable to find their preferred temperature level for a time, they will become more or less active.

The feeding and activity of freshwater fish frequently seems to depend less on the temperature and the state of the weather than on the time of day or the season of the year. The season is perhaps the most important, because it is more closely related to food patterns. *Both, however, are related to temperature.*

At the time of the year when the temperatures are lowest, river and lake fish are often sluggish and feed less vigorously. Unlike warm-blooded creatures, which need food to maintain body temperature, fish need less food when their temperature drops to the lower limits of their tolerance. But they will seek out warmth if it is available, and in cold weather will stay down in the deeper pools, where the water is less likely to be cooled by the upper air. They will move up on sunny days as the surface water warms.

In fresh water the majority of fish appear to be most active when the air temperature is between 60° and 68° F. (15.6° to 20° C.), but even in this temperature range, as soon as the air temperature drops below the water temperature, activity slows down. We can see, therefore, how cold rain or hail might affect fish behavior.

As mentioned earlier, some observers think that the barometric pressure also influences fish activity, but this argument cannot be supported scientifically yet. It does seem true, however, that as the barometer falls, so does the interest of some species of fish in feeding. They seek deeper water, and even there they are likely to be slower to take food. There is some further comment on this phenomenon in Chapter 11.

Since temperature changes with the weather as well as with the time of day, we can see that marine and aquatic animals will certainly respond to really bad weather. Both their feeding and breeding patterns will change.

THE RESPONSE TO TEMPERATURE CHANGES

When prolonged spells of heat or cold result in considerable changes of water temperature, those fish that are most sensitive to such changes will move either to different depths or to other areas. What was said earlier about the correlation between temperature and the size of a fish catch in a season (and this applies also to lobsters, crabs, and other crustaceans on which fish feed) has been confirmed in so many parts of the world that it cannot be ignored by those who fish commercially for our food needs.

The temperature sensitivity of some fish is so acute that they can probably detect a change of as little as a thirtieth of a degree centigrade, certainly a fifteenth of a degree centigrade ($\frac{1}{16}°$ to $\frac{1}{8}°$ F.). As already mentioned, most fish have very definite limits to their temperature tolerance, just a single degree in some of them, but these tolerances change as the fish grow and change in size. In many species the tolerance increases or decreases with the seasons, especially in connection with spawning or when there is a need for a particular diet.

Just as shallow water will heat faster and become warmer than deep water, it will also cool faster and get colder, so a sudden cold snap can set fish moving from one place to another just as effectively as excessive heat. But in the breeding season, some species have quite a change in their tolerance levels, as much as twice their ordinarily preferred temperature range, and this enables them to migrate to particular areas to spawn. They must spawn in places where there is food suitable for newly hatched fry, and this alone controls their need to change their tolerance levels. The swordfish is one species which needs this range of tolerance for spawning.

Although many antarctic marine fish live in temperatures of 30° to 31° F. (−1.1° to −0.6° C.) which is below the freezing point of fresh water, their tissues do not suffer. The fluid in their body cells has precisely the same osmotic pressure as seawater—it is in what is called 'isotonic" state—and has the same low freezing point as seawater.

Some antarctic fish that live in constantly near-freezing water can be cooled to 27.5° F. (−2.5° C) without harm, but they have an upper lethal temperature of 42° F. (5.5° C.), which means that above that temperature they die. Death takes place because of injury to the central nervous system, which is finely tuned to very cold water.

9 / Understanding Ocean Tides

TIDES CHANGE THE TEMPERATURES of coastal waters just as wind and rain do. When they are particularly high, they also affect life well up the rivers.

THE CREATION OF TIDES

Both the moon and the sun share in the creation of tidal movements. The highest tides occur when the moon is new, because then it is on the same side of the earth as the sun. The gravitational pull of both the sun and moon is combined, and the oceans respond to this most powerful attraction.

The moon is a solid body, and therefore much denser than the sun, which mostly consists of burning gases. The moon is also nearer to the earth than the sun. That, combined with the moon's solidity, more than offsets any effect produced by the greater size of the sun. These two factors of density and closeness give the moon 2.34 times as much tidal pull as that exerted by the sun. This figure is arrived at with a simple formula which includes the densities of the three bodies, but we can break it down into two easy sums.

$$\frac{\text{diameter of sun}}{\text{diameter of moon}} = \frac{864{,}000 \text{ miles}}{2{,}160 \text{ miles}} = 400$$

$$\frac{\text{distance of sun from earth}}{\text{distance of moon from earth}} = \frac{93{,}000{,}000 \text{ miles}}{283{,}000 \text{ miles}} = 390 \text{ approx.}$$

The sun's greater distance from the earth virtually cancels out the effect of its greater size. This means that it is the relative densities of the sun and moon that are responsible for the earthly tides.

$$\frac{\text{density of sun}}{\text{density of earth}} = \frac{1}{4} = 0.25$$

$$\frac{\text{density of moon}}{\text{density of earth}} = \frac{1}{1.66} = 0.6$$

The greater effective pull on the moon in approximate figures is:

$$\frac{0.6}{0.25} = 2.4 \quad (\text{The exact figure is 2.34.})$$

Each day the moon reaches a given point on the compass approximately fifty minutes later than it did the preceding day. Therefore, its position at any given hour is 90° farther east every seven days.

When the moon is new, it is on the same side of the earth as the sun, and a maximum tidal effect develops. Seven days later, the moon will be due east when the sun is due north, and there will be 2.34 times the tidal pull (more or less) in the direction of the moon. After another seven days, the moon will be on the opposite side of the earth from the sun (due east when the sun is due west), and so on.

All this suggests that a high tide occurs only once every 24 hours and 50 minutes. But during that time, there are, in fact, two tides. As will be seen in Figure 15, the water also rises on the side which is away from gravitational pull because that side is shielded from the moon's pull by the earth being between it and the moon. But the rise is less than

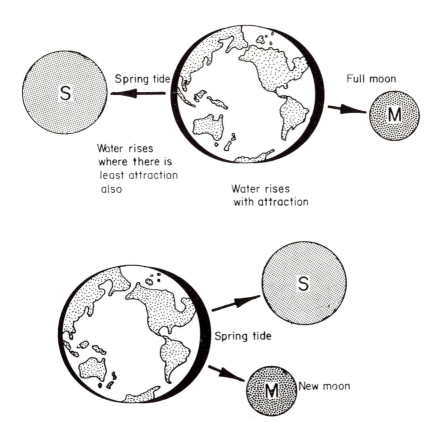

Figure 15 The highest tides occur when the moon and the sun are on the same side of the earth, but even then the tide also rises on the opposite side of the earth. That tide is not as high as the one on the side where the sun and moon have direct pull.

where there *is* pull. Each high tide is followed about six hours later by a low tide.

When a tide is rising, it is said to be *flowing*, and when it is falling, it is said to be *ebbing*. The rising tide itself is called the *flood tide,* and the receding tide is the *ebb tide.* There is about an hour of *slack water* at the low and high points, when the tide is turning.

The timing of the tides is complicated by the fact that although both the sun and the moon create them, the sun exerts its maximum force every twelve hours, whereas the interval between the times when the moon exerts its greatest force increases by twenty-five minutes every twelve hours, or fifty minutes every twenty-four hours. This is complicated further because the sun and the moon are not always the same distance from the earth, so the gravitational pull varies as their distances vary.

Twice a month, the sun and the moon produce the highest tides—once when the sun and the new moon are on the same side of the earth and both exert their gravitational pull, and again when the moon is full and on the opposite side of the earth. Extraordinarily high tides can be expected when the moon is closest to the earth. These high tides are all called *spring tides.*

A *neap tide* occurs when the sun and the moon are at a 90° angle to each other. During neap tides the gravitational pull of the moon and that of the sun tend to counteract each other, and there is a reduced range between high and low tides. Every fourteen days we have spring tides, with neap tides occurring halfway between them.

About every 4,800 years an exceptionally high tide occurs, which can be disastrous in many parts of the world. This happens when the moon is at its closest point to the earth, the earth is at its closest point to the sun, both the sun and the moon are directly over the equator, and the earth, moon, and sun are in a dead straight line.

Don't worry. The next one is not expected until the year 3800 A.D.

But there could have been devastating flood situations around 1000 B.C. and also around 5800 B.C. One wonders if they could account, at least in part, for the floods reported in the mythologies of certain ancient peoples, perhaps even for the biblical flood.

Something similar but not quite so devastating happens

when the moon is nearest to the earth and right over the equator at the same approximate time as the sun. At that time the earth, the moon, and the sun are all in a direct line with each other, *but not necessarily with the earth so close to the sun* as in the previously described situation. At such a time also, the moon and the sun act on the earth together.

A similar but less devastating tide occurred on October 5, 1869. According to a report by the International Oceanographic Foundation, the tide rose 57½ feet in the Bay of Fundy that day. Driven by gales, it rushed inland, wiping out farms and killing thousands of people and cattle. At one place, 121 ships were driven ashore, and one even ended up in a farmyard inland, so the destruction to marine life close to shore must have been most extensive.

The effect of this super tide was very much like that of a tsunami, or tidal wave, which is a giant wave caused by an earthquake, a seaquake, or a volcanic eruption. The size of a tsunami is related to the severity of the disturbance that caused it and the closeness of that disturbance to where the wave strikes.

The explosion of the volcano Krakatoa in 1883 created tsunamis more than a hundred feet high, which inundated and destroyed almost three hundred towns and villages in Java and Sumatra, and smashed three hundred riverboats in faraway Calcutta. Over 36,000 people were killed, so the animal destruction must have been almost without precedent in recorded history.

However, the eruption of Santorin, a volcano in the Aegean Sea, sometime between 1450 and 1480 B.C., is considered to have been several times stronger. The tsunamis created by that catastrophe may have reached speeds of up to four hundred miles an hour. The size of these waves must have been unbelievable, and they are credited with destroying the Minoan cities on the Island of Crete and the Cyclades, and thus with ending the great Minoan Empire. On one of the islands the effects of the waves have been found as high as 820 feet above normal sea level, which gives some idea of just how tremendous they must have been.

Few people realize that lunar phases repeat themselves at the end of every nineteen years. This means, in effect, that the tidal tables for the next nineteen years will be approximately the same as those of the past nineteen years, so that future sea expeditions can be based on tide predictions from an old table. For instance, the year 1976 will have tides that are shown in tables for the year 1957. The only difference will be in the actual height of the tides, which tends to differ even when the time remains the same.

What is more important perhaps is that this knowledge helps in forecasting the recurrence of a devastating tide in areas that have already experienced such tides and the resulting destruction of wildlife and farm stock on land and in river estuaries. But such tide tables will not apply to catastrophes that arose from unusual positions of the sun and/or moon at a particular time. These phenomena do not occur in cycles of nineteen years, but fortunately, as has been described earlier, at much longer intervals.

UNUSUAL TIDES

Not all places have two tides a day. Some places, such as parts of the Pacific, have just one. New Guinea, for instance, has only one high tide a day.

There are places on the north Pacific coast of America where one of the daily tides is higher than the other. Other places even have mixed tides, sometimes one a day, sometimes two. When the moon is exactly over the equator, these places have two almost equal tides, but as the angle of the moon to the equator increases, the second tide gradually disappears.

In Tahiti the tide appears to be controlled by the sun alone, because it occurs at the same time every day. There are other places where the highest tide may not occur on the same day as the new moon or the full moon, but may be as much as seven days later. In some places the high tide does not always occur when the moon is vertically in line with the place. Instead, it occurs with varying amounts of delay, sometimes of several hours; this delay is known as the *lunitidal interval.*

One reason the tides do not behave the same way everywhere in relation to the positions of the moon and sun is because the depth and shape of the ocean affects them. Another reason is the dynamic nature of water itself which is usually moving. Large bodies of land that lie in the flow path also interfere with tidal flow.

Then there is the earth's rotation, which has an effect on tidal currents, just as it does on wind. Tidal currents are deflected to the right in the Northern Hemisphere and to the left in the Southern Hemisphere, the amount of deflection being related to the speed of the current, or flow, so that the deflection is greatest at the poles and nonexistent at the equator.

The crest of a high tide does not move across the oceans in a straight line, but rather in a circular path. In channels between masses of land and in wide river mouths, the banks are too close together to permit a rotary tide, and so quite frequently the tides are different on the two sides of a channel.

Rotary tides have central points around which they revolve. One of these points, which is found to the southeast of New Zealand, produces different tidal times off the coast of that country as the rotary tide flows in an anticlockwise direction from North Island to the south.

RIVER TIDES AND SALINITY

Tides do not stop at river mouths; they proceed upriver to the tidal limits. The resistance a tide encounters in pushing upstream against a current delays the tide in proportion to the distance and the power of the current it must overcome. High tide up a river is therefore never at the same time as on the coast. In fact, it is not unknown for a high tide upriver to be so delayed that it coincides with low tide on the coast. A simple calculation that is fairly accurate is to assume that there will be a time lag of one hour in the upriver tide for every ten miles it must travel.

There are tremendous differences in the rise and fall of a tide in some places, ranging from a barely visible change to as much as fifty feet, and this variation creates very considerable currents as the tide flows into enclosed areas. Strangely, in some areas the strongest currents develop midway between high and low water. These tidal currents are usually quite slow in areas that are open to the sea, but in shallow waters they gather speed and can surge up rivers with great force.

The fresh water from a river naturally mixes with the salty tidal water in its estuary, and both the nature and the movement of this water depend on the relative forces of the current going down the river and the tide going up. When the downstream river current is strong, the heavier salt water in the estuary will form a wedge at the bottom, tapering off upstream. The salt water is always below the fresh, especially after rain.

Often the surface water flows at quite a different rate from the deeper water in these situations, a matter that can be important to certain kinds of fish. When the tidal flow is really powerful, the conflict of movement causes considerable mixing of the salt and fresh waters, and prevents the usual layering from taking place.

The earth's rotation throws the heavier saline water to one side of any large estuary, and this phenomenon is added to the greater specific gravity that carries the salt water below the fresh water. The salt water will bank up on the south sides of large estuaries in the Northern Hemisphere, and on the north sides of estuaries south of the equator.

Another tidal effect found in a few rivers all over the world is known as a *bore*. It is confused by many people with what we have called a tidal wave, but there is a difference. The bore, which is a wall of water flowing upstream, comes with every tide. It happens when the shape of the river mouth obstructs the incoming ocean tide until it builds up into an irresistible force, which then rushes upstream at speeds of as much as fifteen knots, with a frontal wave sometimes several feet in height. Much depends on how high the ocean tide rises, of course. Where it is small, the bore may be only a few inches high.

10 / The Effects of Tides and Salinity on Fish Behavior

FISH FEEDING HABITS AND TIDAL CYCLES

Some fish feed according to the tidal cycle. This in no way changes their temperature preferences. Instead, it reflects their preference for food that appears and disappears with tidal cycles and probably with the temperature changes that accompany a tidal cycle.

In most parts of the world, fish that cling to the coastal belt feed most actively on the flood tide, when smaller organisms and cooler water are washed in during the summer, and warmer water in winter. But there are times when fish feed on the tail end of an ebb tide and the beginning of a flood tide. Each species of tidal-feeding fish arrives with the particular tide phase that its food prefers, whether that phase is the warmer ebbing water giving way to cooler water coming in from below on a sunny day, or the reverse at night.

This may account for the fact that there is a variation within a species from one location to another at times. What may be true of a species in one place may not be true of the same species a hundred miles away. Vagaries of weather that change the temperature of the water can be responsible for this. Also, if the small organisms are swept away by bad weather, the fish will follow.

When a rising tide brings small fish and other marine organisms into shallow waters, bigger fish will naturally follow them and feed on them and so may be present in the area for only a limited period of time. This effect is even more marked if the rising tide coincides with dusk or dawn. There is another aspect to this too. Large numbers of marine organisms, which are food for surface-living fish, move up from deeper waters for the night when the surface water cools. They move down again to the darker depths as daytime approaches and the sun warms the upper layers. These organisms will be washed in by the tide if it coincides with the time they are near the surface.

TIDES IN ESTUARIES

In fresh and estuarine waters there is another factor that may affect the breeding and even the survival of fish, although to what extent we cannot be sure. Any change in the hydrogen-ion content (pH) of the water, which, simply speaking, is an indication of the balance between acid and alkali, will increase or decrease the activity of fish in the area. In estuarine waters tides can play a part in this factor, as does acid runoff from the land to rivers after heavy rain.

A rising tide in an estuary, especially at night, carries in small organisms that have risen to the surface outside. Unable to resist the inflow, they are carried willy-nilly in and out of the estuary with the saline water. That is why many species of fish, both from the sea and from the upper fresh waters, move into the estuaries just after high tide.

RESPONSES TO CHANGES IN SALINITY

So long as the weather is relatively normal, the increase of salinity with the rising tide in an estuary and the subsequent increase of fresh water with the ebbing tide will not be interrupted, but heavy rainstorms over the land will bring down floods of nonsaline, dirty water. Although the salinity tolerance of many ocean fish is considerable, in others it is very critical, so these conditions can be important to the welfare of the latter, and will keep them away from their normal feeding areas.

The importance of salinity to some fish was revealed when controlled experiments showed that some fish can detect a change of as little as one part of salt in two thousand parts of water. That explains why many species simply must move away from an area when the salinity is reduced by surface water runoff. Not all species in estuaries are affected, however. Some have built-in adaptive glands that balance the salt in their bodies by eliminating any excess, and active kidneys that will excrete any excess of non-saline water in them.

When land, salt water, and fresh water meet in one place like an estuary, they produce a variable environment that is very favorable to this kind of fish. This is appreciated by the many sharks that visit estuaries in tropical areas and by the crocodiles living around them. In fact, because estuaries are tidal, the variation in salinity and of inhabitants from microscopic to large creatures can be quite extraordinary from place to place.

In dry tropical and subtropical areas, evaporation from the surface of an estuary may be greater than the volume of fresh water flowing down from the river. If the estuary is small, the salinity concentration of the estuary may even be higher than that of the adjacent ocean. In some places it has been found to be as much as three times that of the sea. The salinity of estuaries in more humid or nontropical areas can also increase through evaporation when there is an unseasonably long rainless period, especially if the river outflow dwindles to no more than a mere trickle. This can drive some species of fish away from an estuary. Even though it is a result of what for us is a welcome long spell of fine weather, the situation proves bad for these fish.

11 / The Effects of Storms on Shore Areas and Estuaries

VIOLENT STORMS AFFECT FISH both directly and indirectly. Whereas milder weather conditions affect only the surface layers of the ocean, storms damage the geographical features and marine inhabitants of both the shore and the offshore regions.

Like the estuaries, these regions are also affected by the vast quantities of storm water that rush down to the sea from flooding rivers or by surface runoff from the adjacent lands—water that carries silt, minerals, and vegetable matter and makes the estuaries and large areas of ocean around them temporarily uninhabitable for countless fish.

This condition will be worse on any strip of coastline with a relatively large number of estuaries. Vast stretches of such coastline can be cleared of all fish which have a salt-extraction metabolism adapted only to a fairly stable salinity of seawater and which need more oxygen than can be found in dirty water. Some fish will be found in such areas, of course, but they will be different species from usual, perhaps scavenging types.

The great quantities of storm-deposited vegetable matter, such as leaves and sea plants torn from their beds, all decompose rapidly and further reduce the oxygen in the water. Soon only fish able to take in air at the surface will get sufficient oxygen—and very few marine fish can do this.

Every living organism lives on some other living organism, whether it is plant or animal. We call this relationship the food-chain link, and the slightest negative effect of climate or weather on any one part of this chain is significant to almost all the rest of the chain right up to man. The 10 percent of the ocean that is on the coast is more important as an animal habitat, and therefore as a source of food for man, than the 90 percent that is open ocean, and it is in the coastal waters that weather has its most devastating effects.

When a sudden storm strikes, fish are often killed in great numbers in shallow coastal waters. They suffocate because the suspended mud and sand in the churned-up water clogs and tears the membranes in their gills and destroys the delicate oxygen-extraction blood vessels. Fish that do survive these conditions are often blinded by the suspended sand, which scratches the delicate surfaces of their eyes.

In September, 1960, a hurricane struck the east coast of the United States. Afterwards, a report given by the International Oceanographic Foundation said:

> Investigations conducted in the shallow waters of Florida Bay after hurricane "Donna" showed heavy kills of many local fishes. Death was usually due to suffocation or erosion of the gills, a result of large quantities of mud churned up by the storm. Oxygen depletion after the hurricane caused an additional massive kill of fish and invertebrates. This depletion resulted from the decomposition of both the organic materials incorporated in the mud and the tons of leaves stripped from mangroves and deposited in the water.

ESTUARY MORTALITY

Fish that spawn in estuaries, and their young which must live in estuaries for a period before moving into the ocean, are destroyed in vast numbers by storms. But the wholesale destruction of fish is not all that happens. The crustaceans and other invertebrates on which so many fish depend for their food are also killed, adding to the local decomposition and oxygen loss. This happens with shellfish, urchins, worms, echinoderms, plants that are torn loose, and countless other things that contribute to the life cycle of the waters. The situation is even worse when an extraordinary low tide coincides with the height of the storm.

Where there is coral, this is damaged too, and the general decomposition and destruction and the consequent absence of oxygen make conditions foul beneath the rocks and ledges. Because the decomposition of everything that dies uses up still more oxygen, other organisms continue to die for some time, thus amplifying an already deadly situation. Fish eggs and fry just dissolve away. So it is easy to see why some areas can be almost completely devoid of life after a great storm and remain so for a long time.

Fortunately, the ocean can cure itself of its own ills, although not of those introduced by man. It eventually becomes clean again after a storm. Lost life, however, can be replaced only by the inward migration of new life from elsewhere, and fish will come to stay only when their basic food needs have been restored. This is the first requirement, and it often takes a long time for microscopic and invertebrate life to return. In some places one to twenty years can elapse before the pattern of life is again what it was before a major weather catastrophe.

Perhaps some sea creatures escape the full fury of a storm by anticipating its arrival in some way we do not understand. Sometimes one place will have a sudden influx of such things as prawns from another area where there have been great rains, but it may not be wise to assume that any real connection exists between the two events.

Once we know just how great the destruction of underwater life can be, the presence or absence of certain kinds of fish in a particular area after a violent storm can be understood more easily. The sudden appearance at such a time of fish that are not normally encountered in the area is rather strange, however. They could be explorers searching for prey or, if they come in large numbers, they could be feeding off creatures that were killed by the storm. They have to be large fish if the water is not yet clear of suspended dirt and decaying matter, because only large fish can escape gill damage from churned-up sand and silt.

MORTALITY IN FISHING BIRDS

Perhaps equal in importance with fish mortality is the damage done to fishing birds by violent storms. They are driven great distances from their customary haunts, sometimes right across the land, where they cannot survive. Often they find themselves in the calm center of the revolving storm system and are carried along to places where they would never otherwise be. Unless they can struggle back to the sea or to large lakes, great numbers of them die.

The nesting sites of shore-nesting birds are destroyed by storms, and young shearwaters become buried in their burrows or sand nests by the sand piling up over them and they suffocate. Some birds are drowned. Those that manage to get into trees or above water and sand may be lifted in updrafts and thrown down a thousand miles or more inland.

Sea- or shorebirds that can get into mangrove swamps, where they will be relatively protected from the wind and driving sand, have a good chance of survival, and may even escape the blinding that they would otherwise suffer from the scratching of their eyes by wind-borne sand—scratches that would rapidly become infected.

Figure 16 A giant wandering albatross *(Diomedea exulans)* was injured and blown ashore in a violent storm off the east coast of Australia. Unable to get back to the sea to feed, it would have died without human intervention.

Figure 17 A young wedge-tailed shearwater *(Puffinus pacificus)* crouching cowed in the sand of an island off the northeastern coast of Australia. Probably millions of these birds die every year in violent storms around Pacific coasts.

We don't really know how many seabirds roam the open oceans, and we never know how many die in great storms, even though we can count thousands of dead seabirds on the shores when calm returns. Undoubtedly, many colonies of seabirds have been founded by groups of survivors blown far from their normal localities in great storms.

All this emphasizes why nature needs time to restore the normal balance of life in an area, and why she sometimes surprises us with the presence of strange species after destructive storms.

THE EFFECT OF STORMS ON MIGRATION THROUGH ESTUARIES

We have already discussed the destructive effects of excessive rain and heavy storms on estuarine fish. But what of the fish that merely pass through estuaries to and from their spawning grounds and must take time to adapt to differences in salinity? Eels and salmon are well-known examples of such fish. The latter especially must delay their transitory period if a storm fills the water with debris and decomposing matter, because salmon must have clean, well-oxygenated water.

Salmon find the way to their spawning grounds by detecting the odor of the stream of their birth. This is quite impossible if the water contains soil and other runoff from massive rain, although it may still be possible after moderate rain or a less destructive storm.

RESPONSE TO CHANGES IN ATMOSPHERIC PRESSURE

The pressure exerted by air averages only 14.7 pounds per square inch of body surface, and even the greatest possible variation in this pressure cannot compare with the variation produced by a change of a foot or two in water depth.

The pressure of water varies from about 15 pounds per

THE EFFECTS OF STORMS ON SHORE AREAS AND ESTUARIES

square inch in the top surface inch to 3.5 tons or maybe 7,850 pounds per square inch in the deepest parts of the ocean. At one foot below the surface it is about 15.5 pounds, and at two feet about 16 pounds to the square inch. When we consider the amount of automatic adaptation necessary when a fish moves down six feet in the water, the small amount of change possible in the atmosphere is not likely to be significant.

When certain fish are brought up suddenly from a considerable depth, they extrude what appears to be their stomachs, and some people have thought that this indicates the reaction of fish to variations of pressure. This extrusion of the internal organs occurs when the gas bladder has not had time to dispose of the high pressure of gas that has slowly built up to counteract the pressure in deep water, and to that extent it is a reaction to pressure variation. However, the small changes that are encountered in the atmosphere do not make this necessary.

The gas bladders of some fish are equipped with openings through which accumulated gas can be expelled rapidly via the mouth or an aperture near the anus. Good examples of this are found in eels, herring, catfish, carp, and pike. But not all gas bladders are like that. Some are sealed, and the gas must be reabsorbed slowly into the bloodstream. It is this kind of gas bladder that is extruded when the fish are surfaced suddenly.

Pressure changes do not affect fish so long as they are not sudden, but storms bring changes in atmospheric pressure that some scientists believe to have an influence on fish behavior. To casual observers, this does appear to be the case, but they could be deceived.

Pressure changes in the atmosphere may affect the food chain through the minute organisms and small fry on which larger fish feed. Fry are exceedingly sensitive to pressure.

PART THREE
HOW LAND ANIMALS ARE AFFECTED

12 / The Effects of Climate Variation on Breeding

THE LENGTH OF A WINTER, any climate variation from the normal, a resulting increase or decrease of available food —all have a direct effect on the breeding habits of many land animals. Among the variations are higher or lower than average temperatures, higher or lower than average rainfall, and an increase or decrease in the number of cloudy days.

Variables of this kind will control the period of hibernation in some species, affect the amount of food available to others, destroy customary breeding grounds by drying them up or, in other cases, flooding them, and upset the balance and production of sex hormones because of interference with day temperature and brightness.

SEVERE AND PROLONGED WINTERS

A severe winter following a poor summer will not only reduce the numbers of some species; it will further affect their breeding behavior because there are fewer of them to breed. Even in areas normally subjected to low winter temperatures, any extension of the winter's duration will cause animals to use up the meager fat reserves they have stored in their bodies during the preceding poor summer. As a result, they may die in large numbers or be taken more easily by predators because of weakness.

During a long winter many seabirds will move inland and compete for the food needed by some of the land animals as well.

Any extended winter condition that keeps to a minimum the number of hours in a day in which the temperature rises to a certain necessary level will produce a semistupor in some birds, bats, rodents, marsupials, and reptiles. This state will be even more marked if food is also short.

A long winter may mean a short summer as well, and so a shorter breeding season for creatures that make up the main diet of other animals, who in turn will also breed less because of limited food for their young. Only a few animals can adapt to this situation, and those that conceive in the autumn and deliver their young in the spring are always affected by lengthened winters.

When spring rains or the melting of snow on mountain ranges is delayed, it may be difficult for animals to find drinking water. In addition, ponds and streams used as breeding grounds by water birds, amphibians, and many insects may prove inadequate. Because some species are tied tightly to a time limit in their breeding, they may miss a year entirely or at least lose a large number of offspring.

Many of the inhabitants of the world's oceans can move in four directions. They can go up and down as well as from place to place, and this gives them a greater chance of escape from drastic weather and climate changes. But except for burrowers and mountain dwellers, land animals can move only in one plane, and this demands great physiological adaptation to ensure survival in times of environmental crisis.

Probably most land animals are bound to their chosen local area of activity—their personal territory, as we may call it. When this fact is added to another—that destruction by bad weather can be greater on land than in all except the surface layers and fringes of the oceans—it is not diffi-

THE EFFECTS OF CLIMATE VARIATION ON BREEDING

cult to understand why animals become scarce in some areas after floods, violent windstorms, and unusual spells of exceptionally hot, cold, dry, or damp weather.

These last conditions can seriously affect the breeding habits of animals, especially those that breed only once a year. For instance, the reproductive glands of many animals are activated by the increasing light as winter turns to spring. Some respond to the actual length of the day as it increases, and others respond to increasing spring temperatures. Quite obviously, an extended winter and consequent shorter spring or summer, or an overcast spring sky, will interfere with the breeding cycle of these animals, perhaps reducing the number of young to carry on the species in the area the following year.

Furthermore, the young require a continuous and plentiful supply of food. When there is any interference with the food chain, such as when weather affects the supply of insects or grubs on which tiny mammals, reptiles, and amphibians feed, fewer of these will survive. Then the animals that prey on these small creatures will have less food, too, and so fewer of their young will survive.

Unseasonable spring weather can make the situation even more difficult for some animals. When the adults require a combination of both a critical temperature and a certain day length, they can be more easily disrupted by weather than when only one of these factors is involved. Cold-blooded animals such as reptiles and fish frequently need this kind of combination. For instance, an American lizard, *Anolis carolinensis,* only responds to the lengthening days if the surrounding air raises its body temperature to at least 90° F. (32° C.). It will not respond normally if the temperature goes above or below this. Because day length and the amount of sun trigger hormones that stimulate breeding in some animals, these animals can be upset by a delayed spring.

Migratory animals returning to breeding grounds will do so in response to different stimulating factors. Birds leave their winter homes with increasing day length, and they may arrive at their summer breeding areas thousands of miles away before the end of severe winter weather if the winter in that part of the world has been extended. This means a lack of the insect food that ordinarily appears in the spring, and so a high mortality rate, or at least delayed breeding, in some species of birds. Delayed breeding means that there is less likelihood that there will be time for a second clutch of eggs if the first is lost by accident or predation. Often it is the ability to repeat egg laying after loss that keeps the population of a species at its normal level.

RESPONSES TO RAIN

In dry areas the breeding capability of many amphibians is stimulated by the sudden appearance of rain. Certain toads, especially in the desert areas of California, breed only when water arrives in the form of spring rains or flash floods. Just how water stimulates the breeding cycle is not quite clear. It may be through increased plant growth or insect life, which meet a basic metabolic need. In some cases, although not in all, it seems to be the very sensation of the rain on the body. There may be a glandular response to water too; because experiments have shown that if the adrenal glands of one species of frog, *Rana cyanophylyctis*, are removed, it will not ovulate or spawn either in rainy or dry conditions.

Quite the opposite happens with squirrel monkeys, *Saimiri sciureus*. These animals live in a tropical area, and the length of day varies by only about eleven minutes throughout the year. Therefore, the length of day does not affect the hormone release that leads to mating. Even though the

THE EFFECTS OF CLIMATE VARIATION ON BREEDING

young are born in the wet season, mating always takes place in the dry season, when the temperament of the males changes completely, and they become more masculine, more vocal, and more aggressive. In this case one wonders if the very dryness of the air stimulates the sex hormones.

13 / The Effects of Temperature on Breeding

ANIMALS, AND THIS INCLUDES birds and insects, are very sensitive to the changes in air temperature, humidity, and air pressure that precede storms, rain, and heat. Even plants react; they droop. Without being fully aware of it, humans also are sensitive to the weather. It is said that business improves with improving weather conditions, and we have all heard people forecast rain because of rheumatic pain in their joints.

ANATOMICAL AND PHYSIOLOGICAL CHANGES

Unseasonal temperatures can affect an animal's breeding habits, but it does not always stop there. In some animals, especially among the lower orders, structural changes take place. For instance, some fish hatch with fewer vertebrae and fin rays after unseasonably cold winters, because the effects of temperature extend to embryonic development. One of the salamanders, *Ambystoma gracile*, also suffers a variation in the number of its vertebral bones when the temperature fails to reach the ideal level while the embryos are developing.

Another effect is seen in the common toad, *Bufo vulgaris*, which produces a much higher proportion of males to females when temperatures stay at about 77° F. (25° C.). The significance of this is not yet understood.

WEATHER AND THE ANIMAL WORLD

Very few cold-blooded animals can do anything about the effects of temperature variation on embryonic development, but pythons are an exception. Observations and measurements have been carried out on the brooding female python, *Python molurus bivittatus*. They show that she not only coils around her eggs, a rare thing in snakes, but she can also regulate her body temperature around them by muscular movement. When the surrounding temperature drops below 91° F. (33° C.), spasmodic muscle contractions start along her body, rather like the shivering of birds and mammals, and this exercise raises her temperature, the amount of increase being related to the frequency of the

Figure 18 The female python *(Python molurus)* is able to raise her temperature by rhythmic muscle contractions similar to shivering, so that the eggs around which she coils will not cool down in a sudden weather change.

contractions. Thus the female python can maintain a body temperature that is as much as 13.1° F. (7.3° C.) above the surrounding air temperature.

Young hatching green turtles are able to use their great sensitivity to high temperature as a protective mechanism. When the temperature rises above 91° F. (33° C.), they remain quite torpid in the egg chamber, then, as soon as the night air cools the sand, they emerge and head for the sea. If they emerged during the day, they would never reach the water before being taken by predatory seabirds. At night a small percentage of them is certain to escape to the open sea.

EFFECTS ON MATING HABITS AND THE YOUNG

The farther from the equator an animal lives, the more frequently is its breeding pattern likely to be disrupted. Temperature varies less in the tropics than anywhere else on earth. The polar bear, for instance, has a highly critical period in which it can conceive, only ten days each year, and if there is no encounter with a male at the right time, which could well happen due to weather conditions, then no cub will be born to her that year.

Although, in the tropics, animals tend to have few disruptions in their breeding patterns, they do have critical breeding periods. The lemur, for instance, has the second-shortest breeding season of all mammals, just two weeks in the year, which means that all young are born at the same time. It would be a catastrophe for the species if the climate were not stable. Similarly, most female wildebeasts, or gnus, deliver their calves during a three-year-week period only. Those born outside this time have less chance of survival.

THE IMPORTANCE OF INSECTS

Creatures that are a nuisance to man react just like those we have more respect for. Insect pests have good and bad seasons too. We don't take much notice of their bad seasons because they bother us less then, but when the spring is really warm and comes early, all these pests hatch early and have a bonanza season. This is usually destructive to man's food supplies and often to other animals too.

When pests have a good season, their chances for survival are often given a further boost by another weather phenomenon. Strong, steady winds not only interfere with bird migration, they also carry insects and spores from place to place. Even the very lofty winds found ten to fifteen thousand feet up probably do this. It is thought that these wind currents have carried black-rust-disease spores from northwest Africa to Britain, where they were brought down eventually in rain. Canadian pollens reach Norway in this way too. Pests such as the Colorado potato beetle, which were formerly thought to have been transported in baggage or merchandise, are now known to arrive at new places borne on the wind.

Although insects can survive great temperature extremes without dying, they do not leave their hibernation locations until the temperature reaches a certain level. Large numbers of insects can survive extremely long, hard winters with subzero temperatures. Some beetles can tolerate temperatures down to $-31°$ F. ($-35°$ C.).

In Europe, the bat is the animal that is most vulnerable to weather changes. Most bats are insect eaters, and their reliance on insects controls their responses to weather. Until insects appear, insect-eating bats dare not use up precious energy reserves they need to keep the heart and metabolism working, so they remain torpid until they can assuage their need for food.

In cold climates bats usually mate in the autumn, before winter hibernation. Males usually die younger than females, so autumn mating may be a safeguard against male deaths in winter. Only in warm countries do bats mate in the spring.

THE EFFECT OF TEMPERATURE ON THE BEHAVIOR OF BEES

Honeybees respond quickly to temperature changes, but their concern is for the developing eggs in the hive. When the temperature is low, they cluster together tightly over the comb in order to keep the incubating eggs in a temperature of 92° to 95° F. (33° to 35° C.). If the temperature goes too high, they vibrate their wings, sending a stream of cooler air through the hive. When the temperature increases to a point where the vibration of their wings is insufficient, they commute back and forth to bring water to the hive or use diluted nectar to close the openings of empty comb cells. Then, by vibrating their wings to make a draft, they evaporate the liquid, and so reduce the hive's temperature.

14 / Special Adaptations

HIBERNATION

Probably the greatest adaptation animals make to adverse climatic conditions, apart from migration to another area, is hibernation, the long period of sleep or torpidity or suspended animation, as it is variously called. When temperatures drop below the level where an animal's food supplies are readily available, it has two alternatives: either to migrate or to hibernate.

Hibernation is possible partly because the animal's bodily tissues require less oxygen as the temperature decreases, and this reduced oxygen requirement conserves energy fuel. In some mammals the amount of glucose in the blood also drops to about one third during winter hibernation. Glucose is the material that supplies energy. If the body can adapt itself temporarily, so that it can get along with less glucose, there is less need for food and for the use of energy in searching for it. This reduction of glucose does not appear to happen in reptiles, but as they are cold-blooded, their energy metabolism may differ from that of mammals.

Torpidity seems to be quite involuntary, beginning with the animals' becoming drowsy, although in many lizards there seems to be a voluntary element; they can lower their body temperature and become torpid at will, burrowing into sand for the period of torpidity. This ensures the conservation of both energy and body water.

Figure 19 The great network of blood vessels in the wing membranes of a bat loses body heat to the surrounding atmosphere. This makes it difficult for the animal to remain warm during hibernation, so it has become adapted to survival with a body temperature like that of the surrounding air in its cave.

Bats in cold climates begin to hibernate in caves when the temperature drops to 50° to 54° F. (10° to 12.2° C.). Ordinarily, they lose up to one third of their weight during hibernation, but an early autumn can cut short the period of activity during which they build up their body fat to sustain them through the winter. In a prolonged winter, when the temperatures remain low and food is absent, the extended hibernation may lower their weight beyond the minimum for survival.

When bats are fully active, their body temperature can be 105° F. (41° C.), but during hibernation it will drop to the same level as the air in their cave, which is usually about 42° F. (5.6° C.)—or even lower in extreme cold. Some measurements of their body heat during this time have shown it to be as low as 29° F. (−1.7° C.). If their temperature drops lower than that, however, they cannot survive, because ice crystals will form in their hearts and lungs.

The bat's phenomenal degree of adaptation is seen in no other warm-blooded animal. This animal's large expanse of wing membrane loses heat to the atmosphere, which is why its body can assume the same temperature as the cave. At the same time the bat's need for oxygen is reduced, so that it uses only about 1 percent of what it consumes during summer activity.

When there are chilly, wet days in the summer, insects are absent, and bats will again become torpid until the temperature rises once more. Thus a wet summer will reduce their opportunities for building up the fat that will help them survive through winter hibernation.

What applies to bats applies similarly to many small burrowing mammals, such as chipmunks and ground squirrels. When winter is prolonged, their stored food, or deposit of body fat, is used up, so that the weaker ones often die before spring breaks. This applies particularly to ground squirrels in the high valleys of the northwestern United States, where snow blankets the ground for up to ten months of the year. The time in which they can feed is especially critical to their survival.

Because birds migrate to warmer climes when winter arrives, one hardly thinks of them as hibernators too, but in fact one or two species do retreat into caves and become torpid, their body temperatures dropping like those of bats, but to a lesser extent. In a sudden cold spell some hummingbirds become torpid, and since these birds have a very high metabolic rate, unseasonable weather can menace them to some extent.

Lesser nighthawks have been known to become torpid as well, and in 1946 Dr. Edmund C. Jaeger and some companions discovered a hibernating poorwill in southeastern California. They kept it under observation for four winters and discovered that its temperature dropped from 106° F. (41.1° C.) to 64.6° F. (18.1° C.). One winter it remained in a state of torpidity for eighty-eight days. This is very rare among birds, and of those mentioned probably only the hummingbird is actually threatened by unseasonal weather changes.

Experiments with mice have shown that not only do they prolong their winter torpid state during a season in which temperatures are lower than usual, but they become torpid even on odd days when the temperature drops unexpectedly. While this protects them to some extent against food shortage, it also reduces the time left to them for breeding in that season and even lessens their ability to breed.

One of the desert pocket mice, *Perognathus formosus,* is almost entirely inactive in midwinter, when it stays below the ground. It does this at other times too, when the temperature drops temporarily. Besides having a close reaction to temperature changes, this species also responds to excessive rain, and all of this limits the breeding period. Because flowers and fruits respond to temperature and rain too, the growth of food for these mice is retarded by adverse weather as well.

REACTIONS TO HEAT AND ARIDITY

For every species of animal that must adapt to cold winters and often to unseasonably low temperatures, there is probably another somewhere else that has exactly the opposite problem, extreme heat and aridity.

Almost all desert-dwelling animals have adapted themselves to conserve their internal water and stores of fat.

They get the water they need from vegetation or prey, and they retain it because their excretion and body evaporation is minimized.

When animals breathe, the exhalation carries with it water vapor from the lungs that is thereby lost to the body. In true desert-dwelling animals this vapor loss is reduced. They do not rely on the cooling process of water evaporation from the tongue, the lungs, or the skin surface to defeat high temperatures. Instead, they burrow or hide away during the hottest times; most are active only at night.

In the animals that do stay above ground through the heat of the day, heat absorption and consequent water loss are often reduced because much of the heat is reflected by their hard-textured, light-colored body surface.

Some desert animals manufacture part of their water needs by a process that uses oxygen from the air and hydrogen from certain foods, the two combining to form water within the body cells. Thus some animals can go months, even years, without drinking in the true sense of the word.

Desert animals have also adapted so that they can accept a great water loss without much distress. For one thing, their kidneys are an important part of their water-conserving mechanism, and they dispose of very little urine. When other animals lose water by perspiration and evaporation, certain salts and urea remain in the body in greater concentration, and if the water loss exceeds a certain level, the salts and urea act as poisons. Some of the body tissues begin to break down; the blood becomes thicker; and its volume is reduced. Because of this, the blood carries less oxygen to the brain and vital organs, and a state of shock results, followed by death.

A 168-pound man can usually accept a water loss that will reduce his weight by twenty pounds before he collapses. A water loss of one-third of this amount, however, can create obvious distress. At some point after he has collapsed, the man will die. Most animals react in exactly the same way, unless they are especially adapted to waterless conditions.

Some frogs are able to survive in arid conditions and can come through long droughts unharmed, because they are able to conserve their body water by voiding their uric acid in almost solid form until they can immerse themselves in water again. Most amphibians in arid country also take shelter to avoid high body evaporation and carry on their main activities at night, but they have made this adaptation over countless thousands of years.

While no animal can live for long if it loses more water than it can obtain or manufacture for itself, many animals can tolerate long periods of starvation because they are able to call on stores of fat built up in times of plenty, when they ate voraciously against lean times. One species of North American gecko, *Coleonyx variegatus v.*, has been found to survive for six to nine months on four days of intensive feeding, which increased body weight up to 50 percent. With ten days' feeding these animals can increase their weight by 80 percent. This fat storage is much less effective, however, with an entire absence of water.

Few of the adaptations to permanently arid conditions are possible for other animals when they are suddenly confronted with a long drought. They can accept a weight loss of perhaps 5 percent due to water shortage, an amount that is possible even in man with diet variation and reduced activity, but ultimately water loss lowers the level of health.

THE GARUMA

Over thousands of years some animals have become adapted to the harshest of conditions. One of these is the garuma, a gull, which eluded man's greatest efforts to discover its breeding place until 1943.

SPECIAL ADAPTATIONS

This most unusual seabird nests far from the sea, on the bare sand in the heart of the Atacama Desert of Chile. Relatively few non-Chilenos have even seen this desert, even though a good highway crosses it. It lies parallel to the coastal desert region in Peru, and to drive the entire length of it takes several days. People in Chile say that no rain has fallen over the area in the whole of recorded history; we know at least that many years can pass without rain there.

The air temperature of the Atacama can be 98.6° F. (37° C.) at night, and as high as 122° F. (50° C.) during the day. At those times the garumas stand over their eggs and act as sunshades, while a slight breeze off the mountains passes beneath them to help cool the eggs. When, as it does, the night temperature drops low, the birds incubate the eggs in the ordinary way. All these precautions are insufficient, however, when unusually high temperatures occur. Then both the birds and their young die in large numbers. With no moisture, the dead birds just dry up and become mummified, hundreds of them blowing around on the sand and stones.

I have crossed the Atacama Desert, and it seems obvious to me that it was once covered by the sea, as were other deserts of the world. Perhaps, although the sea has receded, the garumas have never learned to move their nesting site, but have gradually become adapted to the changing environment as much as they possibly can.

AESTIVATION

There are fish that can survive long droughts which leave them almost entirely, or even entirely, without water. These are the Dipnoi, or lungfishes, of which there are three surviving groups, one in South America (*Lepidosiren paradoxa*), one in Africa (*Protopterus spp.*), and one in Aus-

105

Figure 20 The Atacama Desert in northern Chile, where temperatures sometimes reach 122° F. (50° C.), is the nesting place of a seabird known locally as the garuma (*Larus modestus*). It is a gray gull that flies twenty miles into this six-hundred-mile-long desert to lay its eggs on the bare ground. Many of the birds do not survive the harsh conditions.

SPECIAL ADAPTATIONS

tralia (*Neoceratodus forsteri*). The swim bladders of these fish, which are of ancient lineage, have evolved into useful air-breathing organs.

When the waters in which the Australian lungfish live dry up to small stagnant pools, they use their gills to whatever extent is possible and supplement their oxygen by swimming to the surface to swallow air. On the other hand, the South American and African species use air all the time and extract very little oxygen from the water. These fish also dig burrows in the river mud before it finally dries up and hardens, and then they bury themselves and aestivate, which means they remain torpid or dormant, until water returns, sometimes not until many years later.

The aestivating lungfishes can be dug out in a solid ball of hard mud, and will be quite shriveled if they have been buried a long time. However, after a period of immersion in water, they will fully recover and be as vigorous as before they buried themselves. This is certainly one of the most extraordinary adaptations to excessive drought to be found in nature, and it evolved in these fish millions of years ago.

Figure 21 The Australian lungfish *(Neoceratodus forsteri)* has a well-developed swim bladder lung that enables it to take in air at the water's surface and so survive for long periods in small quantities of stagnant water. Unlike the South American and African species, however, it cannot aestivate completely.

Aestivation also occurs in a number of catfish, which burrow into the mud and await the return of water. This ability is well known in two species, *Saccobranchus fossilis* and *Clarius magur*. Then there are several fish that are able to travel overland in search of fresh pools when their own dry up. These include the climbing perch (*Anabas*), the snakeheads (family, Ophicephalidae), and the cuchia (*Amphipnous*). They are all able to stay on dry land for a period of time.

15 / Extreme Weather Conditions and the Reduction of Animal Populations

DROUGHT

A drought can hardly be called weather when it is a permanent or semipermanent condition, as, for instance, in a desert, which is quite frequently located beneath a permanent high-pressure system. In areas where rain is fairly frequent, however, an interval of just a few weeks without rain can bring about drought conditions quite easily. The life affected by this kind of drought is very different from that existing in natural desert conditions.

Living in a desert environment is not impossible if there is access to some form of water, even if it rains only occasionally. But unseasonal drought conditions not only involve a shortage or complete absence of water but also reduce the number of insects and other natural prey which serve as food for many creatures. The latter, in turn, would normally be the prey of larger birds and animals, so the prolonged absence of rain sets up a chain of attrition that can leave an area almost completely devoid of life.

While desert areas have the worst and the longest waterless conditions, droughts occur in some other places with some regularity, and they have to be met successfully by the animal populations if the species are to survive. Not all droughts occur in hot weather. For instance, the habitual swamplands of wild ducks may dry up in a winter drought, so that the birds are compelled to find waters where there is less concealment for their nests. The young that hatch there in the spring have to spend their early days learning to fend for themselves.

Frequently, water birds must nest on sites that require a long trek back to water and so must face greater hazards from predation by hunting birds, snakes, foxes, and other small carnivores. Worse still, when water birds desert a dried-up area, the next generation will return to nest in the new site. Therefore, even with the restoration of normal conditions in their old nesting region, a long time may pass before the population builds up to full strength again in that area.

When water levels fall in lakes and streams, fish and other water life are crowded into a smaller volume of water. This creates a temporary abundance of food for birds and other animals that can live on fish and crustaceans, but it also ensures their complete or nearly complete annihilation by the hunters and much leaner times to follow. These conditions also encourage the birds to begin nesting earlier because of the plentiful food supply. Then, when ponds and lakes dry up in a bad season, the birds must commute long distances from their nests to other areas to find food for their young. This means fewer feeding sessions, which could retard the growth of some and cause the weakest to succumb as the strongest compete more successfully for the available food.

A not inconsiderable hazard of drought is fire, which can drive small prey animals away from large areas of grassland or forest and kill all those that cannot leave their young or move faster than the fire. Fires can and do start spontaneously at times, but most are due to man's intervention in the natural environment.

Figure 22 Water birds like to nest in secluded and well-covered places like this corner of Lake Titicaca in Peru. When the waters rise to flood level, however, they lose their nesting sites. If a drought makes the waters recede, they cannot nest either, because of the long trek between nesting site and water.

RABBITS ARE NOT INDESTRUCTIBLE

There was a time when Australia, because of its favorable climate, suffered from a plague of rabbits, billions in number, after a few were introduced by early settlers. These rabbits thrived mightily, ate up the pastureland needed by sheep and cattle, and undermined the land with their constant burrowing. Quite an army of men was employed to trap, poison, and shoot tens of thousands of them every day until a virus disease was discovered that brought them down to manageable numbers in a short period of time.

Because of this rabbit plague, Australian scientists studied rabbits' breeding habits intensively, and they discovered that when pasture dried out as a result of drought, the breeding activity of the animals was reduced considerably and, in time, if conditions did not return to normal, stopped altogether. Here we see an instance of unusual weather affecting the breeding habits and population numbers of an animal through its food supply, because that food supply is affected by extreme weather.

DESTRUCTION BY HURRICANES AND TORNADOES

It is easy to see what happens to seabirds that are swept inland or washed ashore in great storms, but we don't see the countless land birds that are blown out to sea to fall exhausted into the water when their strength gives out. Often a migrating group of birds numbering in the millions will be caught by a storm and virtually decimated, especially if it is a hailstorm. As many birds fly down close to the water, they are battered by spray as well as wind, and if they are small birds, few can survive. This accounts for the fact that in some seasons there is a great shortage of certain species of birds in some localities.

The updrafts from storms passing inland lift birds out of trees and dash them to their deaths. Giant waves swamp nesting and roosting places by the shore and drown the occupants. Again, small birds suffer more than large ones, because their very lightness permits them to be wrenched from their holds more easily.

When giant waves sweep inland, they also swamp the homes of burrowing animals and drown all below, even if the water level stays high for only a minute or so. Surface creatures, such as reptiles, small mammals, amphibians, and invertebrates, are swept to sea in large numbers with the receding water.

A tornado is often more destructive in a limited area than any hurricane can be. Fortunately, it is never as extensive as a hurricane, and it will not travel as far. A tornado can hit an animal with a five hundred mile per hour wind and a blow equivalent to a club of several tons' weight, and then leave it in a near vacuum that will burst every one of its internal organs. The following two hundred mile per hour updraft carries the animal into the air, often to a height of three miles, and disintegrates it.

This accounts very readily for the complete disappearance of animals in an area after a tornado has passed through. No trace of them is ever found again, not even of large animals like cattle. Only birds have the ability to evade these fantastic storms, by flying out of their path, but not all of them manage to do so.

FLOODS

Apart from the animals they kill directly, floods from rivers overflowing after excessive rain often have other long-term effects. They destroy burrows and nesting sites, as well as bird courting and parading grounds, thereby delaying or inhibiting breeding. They also cause two rather opposing conditions: erosion, which eliminates tracts of foliage and reduces protective cover, and excess foliage growth at a later date in other places. This excess foliage gives greater protection and cover to prey, so that predatory animals have difficulty feeding their young.

In areas where there are poisonous snakes, large numbers of them get up into trees during a flood and keep people and other animals out of them. In fact, any swampy area where snakes abound is hazardous in a flood, because they will congregate on every bit of dry, raised ground just as other animals try to do.

As does drought, floods affect the animal populations by interfering with their breeding as well as by killing large numbers of them.

One of the most devastating floods known occurred in eastern Australia in January, 1974. It rained continuously and heavily for weeks, some places having twenty-four inches of rain in twenty-four hours, with even more in isolated pockets. Parts of the countryside and some towns were submerged to depths of up to thirty feet. Rivers increased in width up to a hundred miles, and literally hundreds of thousands of square miles of the country were completely inundated. The flood was extended over an area roughly a quarter of the size of the entire United States.

Swimming didn't help many animals. They were just swept out to sea. Those that could get into trees—mostly snakes and a few mammals—were either overtaken by the rising water, starved to death, or died from exhaustion and exposure. No one could worry about this, however; the crisis was so grave that the entire Australian Air Force was occupied with rescuing and feeding isolated and threatened people.

As the floods began to subside and people began to return to their ruined homes, the death toll was calculated. It came to a relatively small number of human lives, but to literally millions of cattle and sheep. Animals, soil, grass, crops were all gone, washed away to sea, where the coastal waters became polluted for great distances. The loss of wildlife was incalculable, because no one ever really knows just how much there is in an area.

Then, in early February, as if nature was really furious with eastern Australia, cyclone Pam struck that very same coast, bringing some of the highest "king" tides ever known. There had been no such combination of disasters in the recorded history of the country. The floods were backed up again; more rain came; coastal areas were destroyed. The pollution and destruction were indescribable, and it was soon obvious that many people would not see their homes again for weeks or months, if ever.

Some of the animals of the area, which were already scarce and protected, will have a long, hard road to recovery. The only ones to benefit were water birds, mosquitoes, and flies. Some water birds raised four or even five clutches of young on the fringes of the floods during the season. The mosquitoes were said to grow so large (they are never small in Australia) that a joke went around that one had dropped onto an airfield and been mistaken for a helicopter until they tried to refuel it.

Australians are always prone to exaggerate the size of their local mosquitoes, but some are, in fact very large, as is to be expected where there are more than a hundred species. Some of them are dangerous disease carriers, and after such serious flooding they breed in billions everywhere. With no cattle to take blood from, they inevitably take it from humans, and that could be a further side effect from the catastrophe. Even before the cyclone struck, cases of mosquito-borne encephalitis were reaching hospitals in the lower reaches of the tributaries entering the Murray River.

Perhaps no other single weather catastrophe could illustrate how devastating such conditions can be to wildlife as well as to man, and how they can change the entire faunal balance over a long term, perhaps even permanently.

Nevertheless, Australians still managed to make light of the whole episode to some extent with their characteristic wry humor. A very old joke was revived, which tells of two mosquitoes that picked up a six-foot man and carried him off to a hill. One said to the other, "Shall we eat him here, or take him home?" The other replied, "Let's eat him here. If we take him home, the bigger mosquitoes will take him away from us."

INDEX

A

Aegean Sea, 67
aestivation, 105, 107, 108
Africa, 96, 105, 107
albatross, wandering, 81
algae, 55
Ambystoma gracile, 93
amphibians, 88, 89, 104, 112
Amphipnous, 108
Anabas, 108
Anolis carolinensis, 89
Antarctic fish, 62
anticyclones, 35
Atacama Desert, 105, 106
atmospheric pressure, 13, 82
atmospheric zones, 17, 18, 19
Australia, 41, 81, 82, 105, 107, 111, 112, 113, 114, 115

B

bats, 88, 96, 100, 101
Bay of Fundy, 67
bear, polar, 95
bee, honey, 97
birds, 80, 81, 82, 88, 90, 101, 109, 110, 111, 112, 113, 115
bore, tidal, 71
breeding habits, 51, 52, 87, 90, 93, 95, 104, 112
Britain, 96
Bufo vulgaris, 93

C

Calcutta, 67
California, 102
Canada, 37
Canadian pollen, 96
carp, 83
catfish, 54, 83, 108
cattle, 11, 114
chickens, 12
Chile, 105, 106
China Sea, 39
chipmunks, 101
Clarius magur, 108
climate change, 51
climate, effects on breeding, 87
climbing perch, 108
clouds
   altostratus, 44, 45
   cirrostratus, 44, 45
   cirrus, 44, 45, 46
   cumulonimbus, 23, 26, 43, 44, 45

cumulus, 23, 43, 44, 45, 46
   nimbostratus, 44, 45
   rooster-tail, 29
   scudders, 44
   stratocumulus, 44
cloudy days, effects of, 55
Coleonyx variegatus, 104
Colorado beetle, 96
crabs, 61
Crete, 68
crustaceans, 61, 79, 110
cuchia, 108
currents
   estuarine, 70
   ocean, 53
Cycladic islands, 68
cycloidal waves, 31
cyclones, 23, 27, 35, 114, 115

D

day brightness, 51
day length, 51, 52, 89, 90
depressions, 35
desert, 102, 203, 209
Diomedea exulans, 81
Dipnoi, 105
drought, 11, 104, 109, 110
ducks, 110

E

earth, rotation of, 71
echinoderms, 79

eels, 82, 83
embryonic development, 60
encephalitis, 115
erosion, 113
estuaries, 13, 68, 70, 71, 74, 75, 77, 78, 82
estuary mortality, 77, 78, 79
Europe, 96

F

field mice, 12
fin rays, 60
fire, 110
fish, 12, 51, 52, 53, 54, 55, 57, 59, 60, 62, 73, 75, 77, 78, 79, 83, 93, 107, 108, 110
  Antarctic, 62
  estuarine, 59, 78, 68
  feeding, 54, 73
  freshwater, 52, 54, 60, 61
  growth, 57, 59, 61, 93
  harvests, 57, 59, 61
  kidney action in, 75, 77
  mortality, 5, 78, 79
  sensitivity, 13, 53, 54, 55, 59, 60, 61, 62, 75, 83, 93
  shoals, 52
fishing birds, 80
floods, 11, 89, 113, 114
food-chain link, 78, 89
foxes, 12, 110
frogs, 104
fronts
  cold, 43, 45
  occluded, 44, 45
  warm, 43, 44, 45

G

gales, 31
garuma, 104, 105, 106
gas bladder, 83, 107
gecko, N. American, 104
glucose, in blood, 99
ground squirrel, 101

H

heat, 17, 62, 102
herring, 83
hibernation, 87, 96, 98, 99, 100, 101, 102
hormones, sex, 51, 59, 87, 89, 91
hummingbirds, 101
hurricanes, 12, 23, 27, 28, 29, 78, 112, 113

I

insects, 88, 89, 93, 95, 96, 101, 109, 115
International Oceanographic Foundation, 67, 78
invertebrates, 51, 112
isobars, 38

J

Jaeger, Dr. Edmund C., 102
Java, 67

K

Krakatoa, 67

L

lakes, 111
land breeze, 20
Larus modestus, 106
lemur, 95
Lepidosiren paradoxa, 105, 107
lesser nighthawk, 102
light
   infrared, 55, 56
   response to, 53, 54, 55, 57
   ultraviolet, 55, 56
lizards, 99
lobsters, 61
lungfish, 105, 107
lunitidal interval, 69

M

Machupicchu, 21
marsupials, 88
mesosphere, 18
mice
   desert, 102
   field, 12
migration, 82, 90, 101
millibars, 38
Minoan Empire, 68
moon, effects of, 17, 63, 64, 65, 66, 67, 69
Murray River, 115

N

Neoceratodus forsteri, 107

New Guinea, 68
New Zealand, 69
Norway, 96

O

Ophicephalidae, 108
owls, 12
oxygen pressure, 13

P

Perognathus formosus, 102
Peru, 21, 111
pests, 11, 12, 95, 96
pH of water, 74
pike, 83
polar bear, 95
pollution, 12, 13, 57, 114
poorwill, 102
prawns, 79
pressure
   atmospheric, 13, 82, 83
   ridge, 39
   systems, 27, 35, 36, 37, 38, 39, 41, 109
   trough, 39
   water, 13, 82, 83
Protopterus spp., 105, 107
Puffinus pacificus, 81
python, 94, 95

R

rabbits, 111, 112

rain, responses to, 13, 51, 90, 91
Rana cyanophylyctis, 90
reptiles, 88, 89, 112
rivers, 70
rodents, 88

S

Saccobranchus fossilis, 108
Saimiri sciureus, 90
salamanders, 93
salinity, changes in, 70, 73, 74, 75
salmon, 55, 82
salt, extraction glands, 75, 77
Santorin, 67
sea breeze, 18, 20, 21, 31
seabirds, 39, 81, 82, 88, 104, 105, 106
seaquake, 12
sediment, 13, 54, 55, 77, 78, 80, 82
sex hormones, 51, 59, 87, 89, 91
shearwaters 80, 81
shellfish, 79
snakeheads, 108
snakes, 110, 113, 114
South America, 21, 105, 106, 107, 111
spawning, 55, 62, 78
squalls, 21
squirrel monkey, 90
storms, in shore areas, 77, 80, 82
stratosphere, 18
Sumatra, 67
sun, effects of, 17, 63, 64, 65, 66, 67
swell, 31, 32, 33
swim bladder. *See* gas bladder

T

Tahiti, 69
temperature, 18, 51, 93, 97
   change, 12, 23, 51, 52, 53, 60, 61, 89, 94, 95, 99, 100, 101, 102
   range, 13, 62, 96
   response to, 52, 59, 60, 61, 62, 89, 93, 94, 95, 97, 100, 101, 102
thermosphere, 17
thunderstorms, 23
tidal
   bore, 71
   cycle, 73
   wave, 67
tide
   ebb, 65, 73, 74
   flood, 65, 73
   king, 66, 114
   neap, 66
   spring, 66
   super, 66, 67
tides
   creation of, 63, 64, 65
   estuarine, 70, 74
   river, 70
   rotary, 69
Titicaca, Lake, 111
toad, 93
tornadoes, 112, 113
trade winds, 23
trochoidal waves, 33
tropics, Cancer and Capricorn, 23
troposphere, 18, 19

trout, 13, 55
tsunami, 67
turtles, green, 95
typhoons, 27, 29

U

United States, 36, 68, 78, 101, 114
urchins, 79

V

vertebrae, development of, 60

W

water pressure, 13, 82
waves
  dimensions, 31, 32, 112
  height, 29, 31, 32
  patterns, 29, 31, 32
weather map, reading a, 38
wedge-tailed shearwater, 81
wildebeeste, 95
wind, 13, 21, 23, 24, 25, 27, 31, 89
  revolving, 35, 37
  velocity, 23–33
winter
  extension of, 52, 87
  length of, 51, 87, 88
  wet, 52, 87
worms, 79